The Reid Lectures on Natural Theology

The Reid Lectures on Natural Theology

Thomas Reid

Editors:
James A. Barham
Jake Akins
William A. Dembski

Inkwell
PRESS

Description
Thomas Reid (1710–1796), a key figure of the Scottish Enlightenment and colleague of David Hume and Adam Smith, succeeded Smith as the University of Glasgow's Chair of Moral Philosophy in 1764. Renowned for his work on epistemology, Reid championed common sense as the foundation of human knowledge. He also significantly contributed to natural theology, which examines what light nature can shed on God's existence and attributes. Reid regularly delivered lectures at Glasgow on this subject, with student notes from 1780 surviving. These lectures, published here, offer insights into the historical interplay of natural science, philosophy, and theology, illuminating today's debate over intelligent design. Excluded from the ten-volume Edinburgh Edition of Reid's works, this revised, thoroughly annotated edition of Reid's lectures corrects errors in prior transcriptions, underscoring their historical importance as well as their relevance to ongoing debates over teleology in nature.

Library Cataloging Data
The Reid Lectures on Natural Theology by Thomas Reid
190 pages, 6 × 9 inches
Library of Congress Control Number: 2025937943
ISBN: 979-8-89946-011-1 (Paperback), 979-8-89946-010-4 (Kindle)
BISAC: REL106000 RELIGION / Religion & Science
BISAC: SCI075000 SCIENCE / Philosophy & Social Aspects
BISAC: PHI008000 PHILOSOPHY / Good & Evil

Book Cover Design:
The cover image was created by Inkwell staff using artificial intelligence.

Inkwell
PRESS

Publisher Information
Inkwell Press, 2321 Sir Barton Way, Suite 140-1032, Lexington, KY 40509
Web: https://inkwell.net
Published in the United States of America on acid-free paper

Table of Contents

Table of Contents

Introduction: Why These Lectures Matter

A Thomas Reid Resurgence

Keith Lehrer begins his intellectual biography of Thomas Reid by asking "How great a philosopher is Reid?" In answer, Lehrer relates the following story concerning his doctoral supervisor Roderick Chisholm:

> When Chisholm was a department chairperson at Brown [University] he received a telephone call from a man saying that he was a busy man but had time to read one serious book in philosophy and wanted to do so. He said that he was not interested in entertainment but simply wanted to read a book with a greater amount of truth than any alternative. Chisholm, wishing to reflect on the matter, said the man should call back the next day, and he would give him his advice. The next day Chisholm recommended that the caller study Reid. It was a sound judgement.[1]

As editors of *The Reid Lectures on Natural Theology*, we think that Chisholm was right, or at least very close to being

---

[1] Keith Lehrer, *Thomas Reid* (New York: Routledge, 1989), p. 1.

right. Moreover, we see in these lectures a timely contribution to ongoing philosophical discussions about design in nature. *The Reid Lectures on Natural Theology*, as presented in this volume, constitute a new edition of manuscript notes originally transcribed in the spring of 1780 by one of his students at the University of Glasgow.

Actually, the edition was new in 2020, when it was published by Erasmus Press under the title *Lectures in Natural Theology*. But Inkwell Press acquired Erasmus Press in 2025, and so these lectures are now being reissued with a new press under a new title. Moreover, these lectures are now folded into a series titled *Inkwell Classics in Evolution and Design.*

We will say more about the manuscript notes to these lectures and its editorial history below, but first a few words about Reid and the contents of the *Lectures*. Thomas Reid (1710–1796) was born in Scotland, in the village of Strachan (pronounced "Strawn"), not very far to the southwest of Aberdeen. He was born in the "manse," the residence of the village's Presbyterian minister, Reid's father. He was educated at primary and secondary schools in Aberdeen, going on to take a degree in divinity at Marischal College, also located in that city, under the supervision of the philosopher and theologian George Turnbull (1698–1748). Upon coming of age in 1731, he was licensed to preach by the Church of Scotland and in due course he received his own ministry in the village of New Machar, north of Aberdeen.

Reid left the active ministry in 1752 to take up a position as lecturer at King's College in Aberdeen, where he helped to form the "Wise Club," a group of professors, ministers, and other prominent citizens, including the theologian George Campbell (1719–1796) and the philosopher James Beattie (1731–1803), both of whom became his lifelong friends. In 1764 Reid was called to occupy the prestigious chair in Moral Philosophy at the University of Glasgow, recently vacated by Adam Smith, a position he held until 1781, when he resigned it in order to concentrate on his writing. A frequent correspondent of the celebrated philosopher David Hume (1711–1776) and the eminent jurist and philosopher Henry Home, Lord Kames (1696–1782), Reid was one of the leading lights of the Scottish Enlightenment. Today, in the eyes of many, he is ranked second in importance among Scottish philosophers only to Hume.

The *Reid Lectures on Natural Theology* have two natural audiences: (1) those interested mainly in Thomas Reid himself, as a thinker of abiding philosophical interest; and (2) those interested primarily in "natural theology," that is, rational inquiry into the question of God's existence and nature. With respect to the first group, Reid's work has undergone a stunning revival of interest since the 1970s, after more than a century and half of neglect—a revival which has only accelerated in the first two decades of the twenty-first century.[2]

Reid is stereotypically presented as the philosopher of "common sense," and as someone principally concerned with epistemological issues. It is of course true that he held that our commonsense view of the world—the way things appear to us pretheoretically—ought always to be the ultimate touchstone

---

[2] A few landmarks include the *Monist* special issue dedicated to "The Philosophy of Thomas Reid" (vol. 61, no. 2) in April, 1978; Hackett's 1983 publication of a low-cost anthology of extracts from Reid's main works: *An Inquiry into the Human Mind on the Principles of Common Sense* (1764), *Essays on the Intellectual Powers of Man* (1785), and *Essays on the Active Powers of Man* (1788); the launching of the journal *Reid Studies* by the University of Aberdeen in 1986; and the publication of Keith Lehrer's *Thomas Reid* in Routledge's "Arguments of the Philosophers" series in 1989 (cited in the opening of this introduction). More recently, the Thomas Reid volume in the "Cambridge Companions to Philosophy" series (edited by Terence Cuneo and René van Woudenberg, 2004) and, above all, the publication of the uniform "Edinburgh Edition of Thomas Reid" in ten volumes (General Editor: Knud Haakonssen, 1995–2020) testify to the present flourishing condition of Reid studies.

of our philosophical theorizing. It is also true that he can be considered a forerunner of such influential twentieth-century thinkers as the British analytical philosopher G.E. Moore (1873–1958) and the American comparative psychologist and proponent of "direct realism," James J. Gibson (1904–1979).

However, it is untrue that the commonsense methodology, or even epistemology more generally, exhaust the interest that Reid holds for contemporary philosophy. In fact, one of his greatest attractions for many thinkers today is the breadth of his interests, extending from logic, metaphysics, and "pneumatology" (philosophy of mind) to the philosophy of nature, ethics, and the philosophy of human society. Reid's background in divinity supplied him with a deep familiarity with the classical (ancient and scholastic) philosophical tradition, which endowed him with an appreciation for the metaphysical aspect of the modern "way of ideas" that was not always apparent in his predecessors Locke and Hume.

It may be argued that the renaissance of interest in Reid goes hand in glove with the renewal of interest in Scholastic philosophy, and in metaphysics generally, that has occurred over the past couple of generations. Reid possessed a virtually unique combination of philosophical virtues, at least among English-speaking thinkers: a penetrating analytical intelligence informed by a serene and candid philosophical personality grounded in a profound appreciation for the experience of all human beings in everyday life, as well as a wealth of knowledge of the history of philosophy.

The Content of the *Reid Lectures*

For readers primarily interested in Reid as a remarkable and singular representative of the Scottish and British philosophical traditions, the *Reid Lectures on Natural Theology* will represent above all a valuable addition to the *corpus* of Reid's writings. For nowhere else does he expound his views on God and religion in such full and often fascinating detail.

Regarding our second prospective group of readers—those interested primarily in the subject matter of the *Lectures*, namely natural theology—a few words on this vast subject will have to suffice. In the *Lectures*, Reid discusses several classical problems relating to the existence of God, of which three in particular are especially worthy of mention here:

1. the cosmological argument;
2. the teleological argument (the argument from design); and
3. the problem of evil.

Let us look at each of these briefly.

The Cosmological Argument

The cosmological argument appears in two main forms: an argument from motion and an argument from the finite temporal existence of the world. The first states that everything that moves must be set in motion by something else; everything in the world is or has been in motion; therefore, the world as a whole must have been set in motion by something outside the world. The second argument is often referred to as the "*kalām* cosmological argument" due to its original development by early, Arabic-speaking theologians known as "*mutakallimūn.*"[3]

---

[3] The term *kalām* means rational discourse about religion; *mutakallimūn* (s. *mutakallim*) are those who practice *kalām*. It is interesting that the term *kalām* is derived from the Arabic root, *k-l-m*, meaning "to speak"; thus, etymologically, the word means something like "debate" or "disputation." Compare the Greek verb λέγω [*legō*], also meaning "to speak," and the nominal forms derived from it, such as λόγος [*logos*]

This argument runs roughly as follows: The world is not eternal, but rather began to exist; everything that begins to exist is brought into existence by something else that already exists; therefore, the world as a whole was brought into existence by something eternal existing outside the world.[4]

Cosmological-type arguments had been discussed regularly down through the ages, right up until Reid's time.[5] On this topic, Reid is largely following expositions advanced by earlier English theologian-philosophers he admired, especially the *Demonstration of the Being and Attributes of God* (1705) by the eminent Anglican minister and friend of Isaac Newton, Samuel Clarke (1675–1729), and the *Analogy of Religion, Natural and Revealed* (1736) by the Anglican bishop and theologian, Joseph Butler (1692–1752).

The Teleological or Design Argument

The second of the three main topics of the *Lectures* mentioned above, and the one to which Reid pays the most lavish attention, is the teleological argument, better known nowadays as the "argument from design." In its essentials, this argument simply states that complex composite objects (such as organisms) whose myriad parts are organized in such a way as to support the well-functioning of the whole (defined as its preser-

---

("word," "speech," "reason") and διαλεκτική [*dialektikē*] ("debate"). This resemblance is no accident, as the historical roots of *kalām* most likely lie in the forms of "dialectic" practiced within seventh- and eighth-century AD, Syriac-speaking, Christian communities in the Near East, who had taken them over from their Greek-speaking co-religionists. See Alexander Treiger, "Origins of *Kalām*," in Sabine Schmidtke, ed., *The Oxford Handbook of Islamic Theology* (Oxford University Press, 2016), pp. 27–43.

[4] See William Lane Craig, *The Kalām Cosmological Argument* (Macmillan Press, 1979; reprinted by Wipf and Stock Publishers, 2000).

[5] See William Lane Craig, *The Cosmological Argument from Plato to Leibniz* (Harper& Row Publishers, 1980; reprinted by Wipf and Stock, 2001).

vation-in-existence) do not appear to be explicable in terms of either natural law or chance (or their combination). Such natural composites would appear to require the same sort of explanation as with manmade composites (like cars or computers)—namely, they seem to require that we posit a mind (or minds) that *designed* them intentionally to mutually support each other so as to achieve a specific purpose.

The *Lectures* offer numerous examples of such teleological phenomena, but perhaps the most intuitively compelling is the case of a *meaningful* text. Here Reid cites the Roman orator Cicero's (106–43 BC) late work *De Natura Deorum* [*On the Nature of the Gods*] (45 BC), in which the author asks whether a "hog grubbing the earth" can form letters making a "complete sentence." Reid claims it is intuitively obvious that such an object cannot be produced by either blind necessity or dumb luck. In our day this is usually cast as the proverbial roomful of monkeys banging away at typewriters to produce Hamlet's soliloquy. By parity of reasoning, the argument goes, no *well-functioning* organism can be produced by either natural law or random chance (or their combination).

This argument was first mooted by Plato (c. 428–c. 348 BC) in the *Timaeus*. There, the Demiurge is invoked to explain the existence of order in nature, especially animals and human beings. After Cicero, the Church Fathers and the medieval Scholastics also returned to this argument with regularity. For instance, Gregory of Nazianzus (c. 329–c. 390 AD), in the second of his five *Theological Orations* (c. 380 AD), compares the world to a lute and its designer, God, to a lutemaker, as well as a luteplayer.[6]

---

[6] Gregory of Nazianzus, "Select Orations," tr. by Charles G. Browne and James E. Swallow, in *A Select Library of the Nicene and Post-Nicene Fathers of the Christian Church, Second Series*, ed. by Philip Schaff and Henry Wace (William B. Eerdmans Publishing Co., 1952) [originally

Closer to Reid's own time, a similar discussion may be found in the *Metaphysicae Synopsis, Ontologiam et Pneumatologiam Complectens* [*Synopsis of Metaphysics, Encompassing Ontology and Pneumatology*] (1742), by the Ulster-born, University of Glasgow-based professor of moral philosophy, Francis Hutcheson (1694–1746). Following discussions of these and similar works, the bulk of the *Lectures* takes the form of the detailed description of innumerable marvels of the natural world, then recently revealed. The whole discussion is organized according to the mineral, plant, and animal "kingdoms," with much attention to the structure of the human body.

One of the most interesting aspects of the *Lectures* for the modern reader is the way it reminds us just how much scientific information—in numerous fields from astronomy, physics, and chemistry to natural history, including anatomy and physiology—was already available to the educated European public by the final decades of the eighteenth century. Moreover, the *Lectures* are also delightful for the way they encourage a sense of wonder in the young minds to which they are everywhere obviously directed.

Reid's discussion of teleology in the *Lectures* remains philosophically significant not just for its contribution to the philosophy of religion but also for its take on the perennial problem of induction in epistemology. Reid delivered the *Lectures* in the year following Hume's posthumous publication of *The Dialogues Concerning Natural Religion* (Hume died in 1776 and his *Dialogues* were published in 1779). And though a friend of Hume's, Reid was not about to let Hume's epistemological critique of the design argument stand.

---

published in 1896], Vol. VII, pp. 185–434. The lute/lutemaker metaphor occurs in Oration 28, "The Second Theological Oration," pp. 288–301, in section VI on p. 290.

Hume had argued in his *Dialogues* that there was no way for induction to justify God's hand in designing nature since we could have no experience of God's activity in nature. As a hard-nosed empiricist, Hume therefore relegated the design argument to the dustbin. Citing Hume by name, Reid countered in Lecture 79 that induction was irrelevant to the design argument, contending instead that we reliably infer design by identifying "the marks of intelligence and wisdom in effects." Reid was thus urging that our minds are hard-wired to see design in marks or patterns that could only be ascribed to "a wise and intelligent cause."

In our own day, Alvin Plantinga has taken this view much further than Reid with the idea of *proper function*, in which our minds, when properly functioning in the right environment, perform various cognitive acts accurately, discerning the truth of the matter.[7] Reid's *Lectures* therefore provide a historically significant counterweight to Hume's thoroughgoing empiricism, which to this day inspires epistemologists and philosophers of science.

Theodicy: Answering the Problem of Evil

The last of the three main topics of the *Lectures* mentioned above is the problem of evil. This is the puzzle regarding how the concept of an all-knowing, all-powerful, and perfectly benevolent deity may be reconciled with the evident existence of evil in the world. It, too, is a problem of great antiquity, originating in its classical form in the various writings of St. Augustine (354–430 AD), and avidly discussed, once again, by

---

[7] Alvin Plantinga, *Warrant and Proper Function* (Oxford University Press, 1993).

[8] Eric Linn Ormsby, Theodicy in Islamic Thought (Princeton University Press, 1984).

the *mutakallimūn*,[8] as well as by numerous ancient, patristic, Scholastic, Reformation, and early-modern authors.[9]

The problem of evil figures prominently in a number of works of Reid's immediate predecessors, notably in *De la Recherche de la Vérité* [*On the Search for Truth*] (1675) by the French Oratorian priest, philosopher, and interlocutor of René Descartes, Nicolas Malebranche (1639–1715); in *De Origine Mali* [*On the Origin of Evil*] (1702; English translation, 1731) by the Anglican Archbishop of Dublin and theologian, William King (1650–1729); and even in the *Essay on Man* (1734), a didactic poem in heroic couplets by the renowned English poet, Alexander Pope (1688–1744).

Above all, the problem of evil is central to the *Essais de Théodicée sur la Bonté de Dieu, la Liberté de l'Homme, et l'Origine du Mal* [*Essays in Theodicy on the Goodness of God, the Liberty of Man, and the Origin of Evil*] (1710) (*Theodicy*, for short) by the great German mathematician, philosopher, historian, and polymath, Gottfried Wilhelm Leibniz (1646–1716). Leibniz, who is frequently cited by Reid in the *Lectures*, argued, not just that the existence of God is compatible with evil, but that the created world we human beings inhabit is "the best of all possible worlds."

To say that ours is the best of all possible worlds does not mean that the world contains no evil. Rather, it meant that any other world that God might have created in its stead would necessarily have been worse than this one, as it must have contained either more evil than this one, or less moral goodness, or both. How is that? Leibniz's argument is complex, but it rests on two principal premises:

---

[9] John Hick, *Evil and the God of Love* (Palgrave Macmillan, 2010 [1st ed., 1966]).

1. a world with free, morally responsible agents is better than a world without them; and
2. even God cannot violate the laws of logic.

It seems to follow from the first premise as a matter of logical necessity that if freedom exists in a world, then the abuse of freedom in the form of evil must be at least a possibility in that world, in which case it is more likely than not that it will become an actuality—that is, that evil will indeed occur. If God grants freedom to humans, then God must have had no choice but to grant them the freedom to do evil, at least if there was to be moral goodness in the world at all.

In other words, God allows evil to exist because it is a logical condition for the existence of moral goodness, and a world of morally good agents is a much better one than a world with nothing but mechanical puppets moving according to the same principles as the planets in their orbits or the balls on a billiard table. If that is so, then any deviation from the fundamental principles of the world as we know it that would reduce evil would reduce moral goodness as well, leaving the world as a whole worse off than it is. And if that is so, then we must indeed inhabit the best of all possible worlds—with the emphasis on the word "possible."

This approach to "theodicy" (a word that Leibniz invented) was often referred to in the seventeenth century as the "rule of the best," or as the philosophy of "optimism" (*optimus* being the Latin word for "best"). It is of some interest that Reid uses a different term. He calls Leibniz's theodicy the "Beltistan theory" (from *beltistos*, the Greek word for "best"), a term that is very rare, but which does occur in a few other sources, notably in the *Lectures on Divinity* (c. 1768) by the Presbyterian minister and early Princeton College president, the Scottish-born John Witherspoon (1723–1794).[10]

Relevance to Biology

So much for the substance of the *Lectures*. Finally, though, we would be remiss if we did not mention yet another reason why this publication of Reid's lectures ought to intrigue perceptive readers in the third decade of the twenty-first century. That is, in recent years, the teleological vision of the living world that Reid's lectures so meticulously document, after suffering a century and a half of intellectual eclipse at the hands of Darwinian reductionism, has once again begun to win a scientific hearing for itself.

In a sense, of course, teleological thinking in biology never really went away. It has been there all along, if only in the form of a subterranean current nourishing biological understanding at a preconscious level for all these decades. What is the evidence for this claim? Biology lectures, biology textbooks—in a word—biologists themselves. It is impossible to listen to any biologist talk for more than a few minutes, or to open any biological textbook to practically any page, without encountering teleological or quasi-teleological (normative, evaluative, intentional, semantic) words and phrases in abundance.

The entire discipline of biology is steeped in teleological concepts and terminology. Take, for example, the following contrasting pairs: "functional/dysfunctional," "healthy/diseased,"

---

[10] The term occurs at the start of Lecture 84 of Reid's *Lectures on Natural Theology* (see especially note 1 of that lecture in the present edition). The Witherspoon example, which appears in the slightly variant form "Beltistian," may be found in *The Works of John Witherspoon, D.D.* (Edinburgh, 1815), in volume VIII on p. 108. We ought perhaps to make clear that while Witherspoon mentions the "Beltistian scheme" in passing, he does not in fact uphold it. In general, he is far less sanguine about rational, or "natural," religion than is either Hutcheson or Reid.

"beneficial/deleterious," or simply "good-for/bad-for." Biologists could not understand or communicate to each other the first thing about their subject if they did not think and speak in such terms—not to mention that conceptual twilight zone which biologists entered some decades ago, where one hears constant reference being made to such utterly baffling notions (from a materialist perspective) as "information," "meaning," "codes," "texts," "messages," "editing," and "proofreading," all spoken with an ontological poker face.

But, wait. Didn't Darwin's theory of natural selection put paid to all of that? Didn't it teach us not to take such ways of speaking seriously, but to regard them as simply a convenient but ultimately expendable legacy effect of an older theistic worldview like Thomas Reid's?

Not really. Or rather, that was the official view given out for public consumption. But the reality was that self-aware biologists always understood the disconnect between their official ideology and their everyday practice. They just adopted the practice of their physicist colleagues who, in quantum mechanics, were also saddled with an incomprehensible theory: "Shut up and calculate!" You didn't need to explain the lingering presence of teleology everywhere in biology in order to do population genetics; you just needed to do the math.

It should have been apparent all along—at least to those philosophers professionally engaged with the life sciences, if not to biologists themselves—that the official story put out for public consumption could not possibly be true, for two main reasons. First, the problem of combinatorial explosion. The more that was learned about the mind-boggling complexity of the cell, the less credible the idea that random variation was the source of evolutionary innovation. Whatever the Bayesian prior of standard neo-Darwinian selection theory may have been 70 years ago, since the advent of molecular biology it has plummeted spectacularly. We now know there is not enough time since the Big Bang for nature to have constructed so much as a single viable protein by means of a random walk, even if biased by selection, through the immensity of the phase space of all possible proteins, viable and non-viable. If evolution occurred, as it seems to many that it did, it certainly did not occur purely through a process of random variation and selection.[11]

Over the past few decades, these things have come to be widely accepted within the mainstream biology community. This is not the place to recount in detail how this came to happen. But, certainly, due honor must be accorded to the thinkers of the "intelligent design (ID)" movement, who laid out the absurdity of the official Darwinian story with logical precision and in exquisite empirical detail. Many of those individuals who engaged in this collective exposé of the state of the Emperor of Biology's wardrobe paid the price for their remarkable stout-heartedness, being vilified and demonized, and having their academic careers derailed.[12] Yet they have the satisfaction of knowing that their sacrifices were not in vain. For, thanks in no small part to the efforts of the ID movement, critics of the mainstream Darwinian view with no such affiliation have for some time now been springing up like mushrooms.[13]

---

[11] Another problem is that organisms are not passive mechanical systems, but are active and adaptable. As a result, whatever the processes of genetic mutation may be, the resulting gene products are actively entrained by the organism into a modified dynamical equilibrium regime, if at all possible. This means that the theory of natural selection presupposes teleology, and so cannot explain it.

[12] Some of the main landmarks in the ID movement are the following: Phillip E. Johnson, *Darwin on Trial* (Regnery Publishing Co., 1991); Michael J. Behe, *Darwin's Black Box* (Free Press, 1996); William A. Dembski, *The Design Inference* (Cambridge University Press, 1998); Stephen C. Meyer, *Darwin's Doubt* (HarperOne, 2014).

[13] Some of the key texts in this area are the following: Franklin M. Harold, *The Way of the Cell* (Oxford University Press, 2001); Mary Jane West-Eberhard, *Developmental Plasticity and Evolution* (Oxford Univer-

For anyone interested in the ID movement and the revolt against the official story it inspired, the *Lectures* will be of great interest. Indeed, Thomas Reid was nothing if not an intelligent-design advocate, even if the term "intelligent design" had yet to be coined. Reid's advocacy of intelligent design centered on the patterns in nature ("the marks of intelligence and wisdom in effects") that he saw as impossible for our minds to understand except as the product of another mind, which for him in this case was the mind of God. Evolution was not in the air in Reid's day, so Reid was not an evolutionist. Yet nothing in his philosophy precludes that the design he saw in nature might nonetheless have gotten there through an evolutionary process. That said, Reid's philosophy is as implacably opposed to Darwin's anti-teleology as it is to Hume's radical empiricism.

This Edition of the *Reid Lectures*

Provenance of the *Lectures*

The text translated below derives from a set of manuscript notes taken by one George Baird, a student who attended lectures on natural theology given by Reid at the University of Glasgow during the 1779–1780 academic year. Baird's notes for these lectures—numbering 119 in all and bound in eight

---

sity Press, 2003); Mae-Wan Ho, *The Rainbow and the Worm*, 3rd ed. (World Scientific, 2008); Stuart A. Kauffman, *A World Beyond Physics* (Oxford University Press, 2019); Marc W. Kirschner and John C. Gerhart, *The Plausibility of Life* (Yale University Press, 2005); Robert B. Laughlin, *A Different Universe* (Basic Books, 2005); Denis Noble, *Dance to the Tune of Life* (Cambridge University Press, 2017); Gerald H. Pollack, *Cells, Gels, and the Engines of Life* (Ebner and Sons, 2001); James A. Shapiro, *Evolution: A View from the 21st Century* (FT Press, 2011); J. Scott Turner, *Purpose and Desire* (HarperOne, 2017). See also the "Third Way of Evolution" website: https://www.thethirdwayofevolution.com.

volumes—are housed today at the University of Glasgow's Mitchell Library under the title "Notes from the Lectures of Dr. Thomas Reid" (MS A104929).[14]

The entire course covered the following topics: the intellectual powers of man; the active powers of man; the human soul; natural theology; theoretical ethics; practical ethics; and politics. The notes treating natural theology correspond to lectures 73 through 87, which took place between February 11 and March 3 of 1780. They are to be found in volumes five and six of the set.

We obtained access to the notes for lectures 73 through 87 through the good offices of Edinburgh University Library's Centre for Research Collections, which possesses photographic reproductions (shelf mark Phot 1211) of the entire set of Reid lecture notes for 1779–1780 housed in the Mitchell Library. The Edinburgh Centre kindly made PDFs of the notes for lectures 73 through 87 available to us.

It should be stressed that these notes are written out in longhand; thus, they are highly unlikely to be notes taken down in class. We know that it was common at the time for professors' lectures to consist of notes read aloud, and for such lectures to be repeated verbatim over the course of several terms or years. As a result of this practice, it is also known that industrious students took shorthand notes, which they later recopied in longhand, with a view to circulating them among their friends or even making them available for sale. Such is the kind of text we are most likely confronted with here.

---

[14] Dale Tuggy, "Reid's Philosophy of Religion," in Terence Cuneo and René van Woudenberg, eds., *The Cambridge Companion to Thomas Reid* (Cambridge University Press, 2004), 28–312 (see, especially, note 2 on p. 306).

Previous Editions of These Lectures

The need for a reliable, annotated text of *The Reid Lectures on Natural Theology* arises from two facts: (1) the *Lectures* have not been included in the now-standard "complete works," *Edinburgh Edition of Thomas Reid*; and (2) although they have been published twice before, there are problems with both editions. Here are the details of the two earlier editions, which we shall refer to below as "Duncan" and "Foster":

- Elmer H. Duncan and William R. Eakin, eds., *Thomas Reid's Lectures on Natural Theology (1780)*. Washington, DC: University Press of America, 1981.
- James J.S. Foster, ed., *Thomas Reid on Religion*. Exeter, UK: Imprint Academic, 2017.

Because neither of these is a proper critical edition, we undertook a careful preliminary comparison of the two editions, which revealed a large number of discrepancies.

Upon examination of the copy of the original manuscript at our disposal, it became clear that Foster sometimes corrected Duncan's mistakes, but that often Foster got wrong what Duncan had gotten right. What is more, in a comparatively small number of cases we felt compelled to advance a third reading of the manuscript, when both Duncan and Foster were clearly in error. It was this fact, more than anything, which convinced us that a new edition was desirable.

Not all mistakes and omissions in Duncan and Foster were negligible. Some were substantive and are corrected in our edition. For instance, we have adverted to the "Beltistan theory," which derives etymologically from the Greek word for "best," and refers to Leibniz's theodicy as this being the best of all possible worlds that God could have created. Neither Duncan nor Foster shed any light on this terminology or how it connects to Leibniz's theodicy. Our edition clarifies this and other problems with Duncan and Foster.

Accordingly, we have reviewed the entire manuscript and established what we are confident is the best existing version of the text (though admittedly still not a proper critical edition). We have also corrected a number of mistakes found in the informational footnotes of both Duncan and Foster, and added a large number of new notes of our own for the reader's convenience.

Editorial Conventions

Finally, a few words concerning the conventions we have adopted in translating the text below:

- *Spelling*—Baird's spelling has not been modernized, though to a limited extent it has been regularized. Thus "antient" appears instead of "ancient" and "shew" instead of "show."
- *Capitalization*—Baird often capitalizes abstract nouns and noun phrases, especially those associated with the deity, though he is far from consistent. We have regularized his practice locally in a few instances (e.g., in analogous phrases occurring in close succession) without attempting to impose global consistency.
- *Paragraphing*—The original text has very few paragraph breaks; for readability, it was deemed necessary to divide the resulting large blocks of text. This has been done primarily with an eye on the sense, while keeping the size of the resulting paragraphs approximately the same to the extent possible.
- *Punctuation*—Our punctuation of Baird's text is a compromise between respect for what he wrote and

accessibility to a twenty-first-century reader. Since Baird himself is very far from consistent in his use of punctuation marks, we have not attempted to impose an artificial consistency on the overall text. On the other hand, following the manuscript's punctuation slavishly would have resulted in a text more difficult than necessary for modern readers to construe. For these reasons, we have punctuated the text with an eye to readability, without modernizing just for the sake of modernizing. In some places, the punctuation may still appear strange to the modern reader. However, we believe the compromise we have arrived at ought to facilitate the reader's comprehension without introducing a grossly anachronistic note into the text.

- *Italics*—Not infrequently, Baird emphasizes common nouns and phrases by underlining them. We have set the corresponding text in italics. In other instances, such as book titles, where the modern reader would expect italics, Baird does not use them, and neither do we.

- *Difficult Readings and Omissions*—In some places, we have been unable to establish with certainty what Baird wrote in the manuscript. In other places, he has omitted words clearly required by the sense of the text. In both of these kinds of instances, we have supplied what we believe to be most likely the correct reading in brackets [like this]. In a few especially difficult or interesting cases, we have provided additional information about the textual problem in a footnote.

- *Explanatory Footnotes*—The manuscript makes lavish use of allusions to classical and contemporary philosophers and other authors. Reid could assume a level of general culture in his student audience that has long since disappeared from the modern world. Thus, annotation of the text seemed indispensable. As it would have been arbitrary to assume that some famous names (say, Socrates) required annotation, while others (say, Caesar) did not, we have provided brief explanatory footnotes for all the proper names, book titles, direct quotations, and obvious literary allusions occurring in the manuscript.

- *Lecture Titles and Summaries*—At the start of each lecture, we give a descriptive title to the lecture and include a summary of it. These titles and summaries are of our own doing and in no way part of the original lectures. We included them because we thought readers would find them helpful in quickly orienting themselves to the individual lectures.

- *Direct Quotations*—In several places, Baird records what are ostensibly verbatim quotations from various classical sources, sometimes in foreign languages. Almost invariably, he gets them wrong to one degree or another. Whether this is entirely Baird's own fault, or the blame is to be shared with Reid himself, is impossible for us to say. In any case, we have located the original source of each such quotation and inserted the authentic wording into the main text, on the assumption that the details of Baird's mistakes will be of limited interest to readers. If the quotation is in a foreign language, we have also provided an English translation in a footnote (unless one has already been provided in the body of the text, as sometimes occurs). For quotations in Greek, we also provide a transliteration into the Latin alphabet.

- *Readability*—Although we have tried to stay faithful to the original notes of the *Lectures*, at times there was nothing to be gained by staying faithful, and indeed something to be lost, when the original clearly impeded readability. Thus, occasionally, we preferred readability over faithfulness. For instance, when the original lecture notes read "twice three will not make 6," we prefer "twice three will not make six." Or when the philosopher Descartes is referred to as "des Cartes," we prefer the usual spelling. Or when points are numbered, as they often are in the *Lectures*, we prefer 1) for point one, 2) for points two, etc. rather than, as in the original, 1. for point one, 2. for point two, etc. Such changes to enhance readability are non-invasive and rare, but on balance seemed to improve this edition.

Background to This Edition

As a theology student at Princeton Theological Seminary in the mid-1990s, Bill Dembski encountered Elmer Duncan's edition of Thomas Reid's 1780 lectures on natural theology while perusing the stacks at the seminary's library. Princeton's theological collection at the time was second in the U.S. only to Union Seminary's in New York.

Duncan had published the lectures with University Press of America as essentially a mimeographed typescript. It was expensive, shortly to go out of print, not pleasantly readable, and with many errors and gaps. Nonetheless, Bill instantly saw the relevance and importance of these lectures for the credibility and intellectual vitality of the intelligent design movement.

Bill therefore determined to do another edition that eliminated the faults of the Duncan edition while also making it readily available to interested readers. As a philosophy professor at Southwestern Seminary in Ft. Worth in the 2000s, Bill had Jake Akins as a student. Bill was so impressed with Jake that he recruited him to become his research assistant.

Jake was tasked with getting a PDF scan of the Reid lectures from Glasgow's Mitchell Library, scrutinizing it word for word and letter by letter, and thereby correcting and upgrading the Duncan edition. In the process, Bill and Jake not only improved on the work of Duncan and Duncan's own student assistant, William Eakin, but also had a pleasant lunch with Elmer Duncan himself, who offered helpful insight and gave his blessing to the new edition (Duncan died in 2016).

Despite all this groundwork, in 2012 Bill left Southwestern Seminary, and the project stalled. For seven years it stayed in cold storage until, in 2019, Bill's good friend and colleague James Barham took, at Bill's urging, the project out of cold storage to assess where it stood and whether it might be fruitfully rekindled.

In the meantime (2017), James Foster had published his own edition of the Reid lectures (with a foreword by Nicholas Wolterstorff, no less). James initially was skeptical that the work by Bill and Jake, even if suitably updated and upgraded, could meaningfully improve on Foster's edition. Nonetheless, it quickly became clear to James that even though Foster had gotten some things right that Duncan had gotten wrong, Bill and Jake's work offered significant improvements over both Duncan and Foster.

James therefore painstakingly went through the editions of Duncan and Foster, the manuscript and research work of Jake, and the original Reid lectures as provided electronically by Glasgow's Mitchell Library, striving to take the best from each and thereby produce the strongest edition of the Reid lectures to date. Barham also added extensive footnotes to help contemporary readers who might not be familiar with the classical allusions rampant throughout the Reid lectures. Toward the end of this project, Cameron Wybrow, himself a classicist, offered insights to improve this edition, in addition to serving as a proofreader.

These lectures were published in 2020 by Erasmus Press and titled simply *Lectures on Natural Theology*, with Thomas Reid as the author, and with only James and Jake listed as editors. They had, after all, done the heavy lifting on this project. Nonetheless, when Inkwell Press acquired Erasmus Press in 2025 and set about reissuing these lectures, it seemed appropriate to credit Bill's role on this project by adding him as a third editor.

Under Inkwell Press, this edition of the *Reid Lectures* is now available not only in hardcopy but also in ebook, making it more widely available in this increasingly digital age. Finally, the editors would like to thank Elmer Duncan, James Foster, and Cameron Wybrow for their contributions that helped bring this project to life.

James A. Barham
Jake Akins
William A. Dembski

Inkwell Classics in Evolution and Design

Inkwell Classics in Evolution and Design is an extended series of classic texts (books, monographs, and anthologies) on evolution and design—as well as related topics—that Inkwell Press will publish in coming years. To hear only scientific naturalists or materialists since the time of Darwin, one would think that non-teleological views of evolution are alone compatible with the progress of science. Thus, any attempt to give credence to teleology or design in biology is said no longer to be tenable.

But in fact, design has always been part of the discussion about biological origins. This series will give equal time to the main voices in this discussion. We will thus publish Darwin, Huxley, and Haeckel. But we will also publish Paley, Babbage, and Wallace. Nor will this series confine itself to texts written before the twentieth century. A serious, intellectually vital debate about evolution and design has existed not just since the time of Darwin. It continues to the present. It has also existed since antiquity, as witnessed by Greek atomists such as Democritus and Epicurus versus Greek teleologists such as Plato and Aristotle.

As for the intellectual merits of this debate, it matters not whether at various periods one side or the other has dominated the limelight. Nor is the question whether some sort of evolutionary process may be responsible for life's emergence and subsequent development. The question is whether purely material or physical forces, lacking any inherent teleology or design, can bring about the complexity and diversity of living forms. It is the key question about the nature of nature.

This series aims to restore a historically honest balance to the debate over evolution and design in the study of biological origins. Many of its texts will focus specifically on biology. Others will deal with philosophical or scientific issues broadly relevant to questions about biological evolution and design. The present lectures by Thomas Reid fall in this category.

Lecture 73: Reason and Revelation

Summary: Man alone among creatures can know his Creator, and such knowledge elevates the mind and sustains virtue. True piety depends on right conceptions of God's nature and providence. Although revelation confirms and extends the truths of natural religion, reason remains necessary to judge its authenticity and interpret its meaning; without reason, revelation degenerates into superstition or fanaticism. Human understanding of God is necessarily analogical, drawn from reflection on the operations of our own minds. Reid adopts Francis Hutcheson's threefold plan for natural theology: to consider (1) the existence of God, (2) His nature and attributes, and (3) His works. Against speculative atheism, Reid identifies two chief causes: false philosophical systems that explain the world without intelligence, and moral motives to escape accountability. He refutes atheistic inferences that deny a future state, moral order, or happiness under providence, showing that even on atheistic premises, the soul's survival and moral consequences remain probable. Belief in divine governance, by contrast, provides comfort, moral restraint, and social order. Finally, Reid affirms the necessity of a first cause: everything that begins to exist must have a cause, and an infinite regress is impossible, thus establishing the existence of an eternal, uncaused, and necessary being.

Of all the animals which God has made it is the prerogative of Man alone to know his Maker. There is no kind of knowledge that tends so much to elevate the Mind as the knowledge of God. Duty to God forms an important part of our duty and it is the support of every virtue; it gives us magnanimity, fortitude and tranquility; it inspires with hope in the most adverse circumstances, and there can be no rational piety without just notions of the perfections and providence of God. It is no doubt true that Revelation exhibits all the truths of Natural Religion, but it is no less true that reason must be employed to judge of that revelation; whether it comes from God. Both are great lights given to us by the Father of Light and we ought not to put out the one in order to use the other. Revelation is of use to enlighten us with regard to the use of Natural Religion. As one Man may enlighten another in things that it was impossible could be discovered by him, it is easy then to conceive that God could enlighten Man. And that he has done so is evident from a comparison of the doctrines of Scripture with the systems of the most refined heathens.

We acknowledge then that men are indebted to revelation in the matter of Natural Religion but this is no reason why we should not also use our reason here. Revelation was given us not to hinder the exercise of our reasoning powers but to aid and assist them. It is by reason that we must judge whether that Revelation be really so; it is by reason that we must judge of the meaning of what is revealed; and it is by Reason that we must guard against any impious, inconsistent or absurd interpretation of that revelation. As the best things may be abused, so when we lay aside the exercise of reason Revelation becomes the tool of low Superstition or of wild fanaticism and that man is best prepared for the study and practice of the revealed Religion who has previously acquired just Sentiments of the Natural.

The best notions of the divine nature which we can form are imperfect and inadequate and are all drawn from what we know of our own Mind. We cannot form an idea of any attribute intellectual or moral as belonging to the deity, of which there is not some faint resemblance or image in ourselves. As we cannot form the least conception of Material objects but must somehow or other resemble those we perceive by our senses so our knowledge of Deity is grounded on our knowledge of the human Mind. And for this reason I thought it best to give you a view of it before we entered upon this subject. In speaking of Natural Religion I shall adopt the plan which has been followed by Mr. [Hutcheson][1] in a tract which he has published and which I shall take this opportunity of recommending to your attention and careful perusal. The first branch is to treat of the Existence of God, 2) of his Nature and attributes, 3) of his Works.

1) The existence of the Supreme Being is so loudly proclaimed by everything in Heaven and Earth, by the structure of our own bodies and the no less curious structure of our Minds and indeed by everything about us, that it may perhaps appear unnecessary to confirm a truth so evident. But when

---

[1] Francis Hutcheson (1694–1746), Scottish philosopher and prolific author, especially known for his works on ethics and aesthetics. The brackets indicate the fact that Baird wrote "Hutchinson," not "Hutcheson." Duncan (p. 11, n. 1) speculates that the "tract" in question is *Moses's Principia* (1724) by the English theologian and philosopher, John Hutchinson (1674–1737). However, nothing in *Moses's Principia* accords with the threefold plan mentioned here by Reid, whereas—as Foster points out (p. 30, n. 3)—Hutcheson's *Metaphysicae Synopsis, Ontologiam et Pneumatologiam Complectens* [Synopsis of Metaphysics, Compassing Ontology and Pneumatology] (1742) contains a section on natural theology ("De Deo" [On God]) that does follow just this scheme. Moreover, other contemporary authors are on record as misspelling Hutcheson's name ("Hutchinson" being the *lectio facilior*). For example, John Witherspoon (1723–1794) made the same mistake as Baird (and Reid?) in his *Lectures on Moral Philosophy*, written in 1768 for delivery to Princeton College undergraduates (see Thomas P. Miller, "Introduction," in *The Selected Writings of John Witherspoon*, edited by Thomas P. Miller [Southern Illinois University Press, 1990; paperback edition, 2015], p. 36).

we consider the importance and that there have not been
wanting persons who have exercised their wits to weaken its
evidence, it will appear proper to consider the grounds on
which it is [suspected][2] and to inquire into the force of the
sophistical arguments that have been urged against it. I shall
therefore point out some observations that appear to have the
greatest strength in confirming this important truth. I shall
first however offer a few remarks on the causes of speculative
Atheism and consider if it can justly be drawn from this system
that there is no God.

I conceive then that there are chiefly two causes that may
be assigned for the Speculative Atheism that has appeared in
the world. There were a few among the antients that professed
atheism as Diagoras,[3] Theodorus[4] and Protagoras[5] and in later
times we are told that Julius Caesar Vaninus[6] suffered death
for atheism in the dark ages. What seems to have led them to
embrace their opinion may, as I said, be ascribed to two causes:

1) To false systems of philosophy by which they thought to
account for the formation of the World and what happens in

---

[2] Another possible reading is "supported," but "suspected" accords
better with the sense.

[3] Diagoras of Melos, fifth century BC, Greek Sophist (an educator/
orator practiced in debate, who was willing to argue any side of a question;
for this reason, Sophists were often reproached for being indifferent to
the truth).

[4] Theodorus of Cyrene, late fifth century BC, depicted as an associate
of Protagoras (see following note) in Plato's dialogue, *The Sophist* (see
note 2 of Lecture 83 below).

[5] Protagoras of Abdera, c. 490–c. 420 BC, credited by Plato (see note
2 of Lecture 83 below) with being the originator of the Sophist movement.

[6] Lucilio Vanini (1585–1619), an Italian freethinker who wrote under
the pen name "Giulio Cesare" [Julius Caesar]. Author of the notorious
work, *De Admirandis Naturae* [*On the Marvels of Nature*] (1616), Vanini
was arrested in Toulouse, where his tongue was pulled out with pliers and
he was burned at the stake. Note what Reid considers to be the "dark
ages"!

it without once bringing in a wise and intelligent maker. They conceive that by a mixture of moisture and drought the mighty machine of the Universe was provided without any intelligence to begin, regulate or finish its operation. The philosophers of the Ionic School[7] were generally thought to lean to Atheism, because that philosophy was chiefly employed in accounting for the formation of the Universe. Everything arose from chaos by a mixture of the elements and we find that Anaxagoras was the first who thought it necessary to introduce Mind into the system and who thought intelligence necessary to put all things in order. All other antients who differ from Anaxagoras must either hold that the world existed from all eternity without a cause or was produced without an intelligent cause and author. The ignorance of true philosophy which leads men to discern marks of wisdom and design in the formation and government of things may be considered then as one cause of Speculative Atheism. But,

2) It was intended by some to free men's minds from the fear of punishment for their crimes in an after state to free them from all reflections on the future or remorse for the past. Epicurus[8] who does not deny the existence of God, but thinks he does not interest himself in the affairs of the World, glories in

---

[7] "Ionia" refers to the eastern shore of the Aegean Sea, now a part of Turkey but then settled by Greeks. The "Ionian school" of natural philosophers (*physiologoi*) included Thales (c. 620–c. 546 BC), Anaximander (c. 610–c. 546 BC), and Anaximenes (c. 586–526 BC), all from the town of Miletus; Xenophanes of Colophon (c. 570–c. 475 BC); Heraclitus of Ephesus (c. 535–475 BC); Anaxagoras of Clazomenae (c. 510–428 BC); and Leucippus of Miletus (mid–fifth century BC).

[8] Epicurus of Samos (341–270 BC), founder of the school known as the "Garden," or "Epicureanism," maintained that the gods are indifferent to human beings, and the aim of life ought to be the pursuit of pleasure ("hedonism"); he was a follower of atomism (see n. 14 below), but added his own idea of the *klinamen*, a spontaneous "swerve" in the motion of the atoms, thus avoiding hard determinism.

this as a great benefit done to Mankind, that he had freed them from the fears of Religion and all the evils which a dread of the gods never fail[s] to create. Lucretius celebrates his praises on this account and by all his colleagues he was also reckoned a kind of deity, who had nobly delivered them from the bugbears of Religion. He seems to have taken it for granted that as there was no Supreme Being, that therefore there was no future life, nothing after death, neither rewards for Virtue nor punishment for Vice and of consequence they might pursue their pleasures without any restraint. This was the conclusion, supposing the principle true, which they drew; but I conceive that these men have reasoned ill even on the principles of Atheism, and before I enter on the arguments in proof of the existence of deity, let us consider the conclusions that follow from this system of Atheism and whether their conclusions justly can be drawn from it, on account of which it seems to have been adopted by some.

It is indeed difficult to reason on a hypothesis so absurd, but let us for a moment suppose that Heaven and Earth are either from all eternity without any cause or were produced by chance without the interposition of any intelligent power. Suppose then that this is the case, will it follow from thence 1) that men must perish at death and that there is no future existence, 2) that if there is a future life that it will have no relation to the present and that our happiness and misery in it will not depend on our conduct here, 3) whether it would tend to make us happier in the present life if there were no God, no future state, and all things governed by unalterable laws, than under the persuasion of a universal ruler who governed all things wisely and well. These it may be proper to consider a little and see whether they may justly be deduced from the principles of Atheism or not.

1) There is no Supreme Being, therefore there is no future state and men must perish at death. Now to me there does not appear the least shadow of connection between the two propositions. If our present existence is consistent with his nonexistence why not also a future? That the thinking principle is distinct from the Body which we see and feel has already been proved by convincing arguments; now whatever that principle is, let us suppose it produced by Necessity or Fate or any other unmeaning name you please to give it; why may not that Chance which at first united these organs disunite them? We see other unions broken which appear equally strict, as that between the Mother and the Child, the Egg and the Bird. Thus it appears that even on the principles of atheism there is not the shadow of evidence for the position; we may grant the premise yet the conclusion will not follow. But further we may observe that in the whole course of Nature we have no proof of the annihilation of any one substance that exists. All the operations of Nature consist either of the composition or decomposition of what already exists without either creation or annihilation.

But the Soul is a subtle simple principle and therefore can perish only by annihilation of which we have no evidence in all Nature, and it cannot reasonably then be supposed in this instance. This argument is equally strong even on the supposition of Atheism because it is drawn only from what we observe of the course of Nature. We may observe, too, that it is agreeable to the Analogy of Nature that we should pass through different states as different from one another as the present from the future. Neither is this grounded on the supposition of a deity, but retains its full force even allowingthe principle of Atheism to be true. We see then that the Atheist cannot solace himself in this conclusion that because there is no God therefore there is no future state; he reasons ill even from the principles of Atheism.

2) The next conclusion I numbered was if there is a future state, will it have no relation to the present and will our happiness or misery there not depend on our conduct here? And here again I maintain that this will not follow. We have already seen that granting the principles of Atheism to be true yet we cannot conclude from them that the *Soul* will not exist after death. We have seen that there are no good arguments against a future state, nay that there are some arguments in proof of it which retain their force even on the supposition of Atheism being just.

Let us now then suppose a future state without any Supreme intelligent ruler, then on that supposition let us inquire whether it is certain that wicked men will not be miserable there or if we will not find our good behaviour here redound to our happiness hereafter. I answer, that it is neither certain nor probable that it will not be so. Nor does any system of atheism furnish us with any satisfying evidence that we will not reap the fruit of our doings. Nay it is probable on the contrary that we shall for, 1) Where animals pass through different states, the future always has a connection with that which went before. Thus if a chicken in the Egg receives any blemish it always retains it and if it loses a foot or a wing it never after recovers it. In like manner a child in the womb if it brings any defect or disease along with it, it continues through life and often renders its days few and evil. It is analogous then that we should carry our good or bad habits along with us to a future world. That vice is a disorder of the Mind is as evident as that lameness is of the Body; now, we have no evidence that it shall not be continued in a future as it is in the present life.

The excellence and superiority of temperance, prudence and fortitude above their contrary vices is intrinsic and results from the Nature of Virtue and Vice and we may as well suppose that twice three will not make six, as that there will be no distinction between them hereafter. If therefore we reason from analogy we see that there is a probability of a good man's carrying along with him the fruit of his virtuous improvement of his rational and moral powers. These are his most valuable acquisitions and if death doth not put a period to his existence we have no reason to think that it will put a period to them. The vicious man has the same probability of feeling the consequences of his bad habits. But, has the Epicurean[9] any probability of finding those objects to gratify his sensual pleasures? Has the votary of ambition any probability to hope that his power will go along with him, or that there he may raise himself to influence and gratify his lust of dominion? Or has the covetous man reason to think that he will carry his idol along with him or that he will draw bills of exchange of this World? So it appears then, on the supposition of a future state, even with- out a Supreme intelligent ruler, that every probability promises happiness to the Good and misery to the Wicked. We see that it is in the course of things here and we can only judge of the future by the past.

Were we to reason farther on this subject, it would not be difficult to shew, that whether we suppose the future world to be a social or solitary state, and if social, whether mixed of the good and the bad, or if we suppose the good and the bad separated, that in all of these cases still the chance is on the side of the Good. We have no reason then but to think that Virtue and Vice will always retain their Nature and produce their consequences. Let a man habituated to sloth and rapacity and injustice, change his climate or his country, let him live in regions of savage rude- ness or in more polished society, let him leave even the converse of men and try the life of a hermit; still will he find the bad effects of his habits follow him to the court

---

[9] Follower of Epicurus; see note 8 of this lecture above.

or the camp, the city or the desert. From the torrid to the frozen zone, he will find his habits noxious and sooner indeed may the Ethiopian change his skin than a man by altering his condition change his habits. What reason then has he to suppose that a passage to futurity shall wash away all his stains?

Thus have I shewn that even supposing there was no God, yet that this affords no argument against a future state, or why our conduct here should not affect our condition hereafter. I come to consider the

3) Conclusion whether men would be happier in the present life if the belief of a God or a future state were removed than under the impressions of a wise and righteous maker and governor of the World. I think it unfair here to compare, as Mr. Bayle[10] has done, the consequences of Atheism with those of the lowest kinds of Superstition, as I plead not the cause of Superstition but of Religion. Suppose it true, however, as perhaps it may be, that these may create some notions even more pernicious than what will follow from Atheism, yet this is not to the purpose.

The abuses of the best things are always the worst. Men by abusing their reason depress themselves below the level of brutes. The abuse of meat and drinks is attended with hurtful consequences; so is it also with religion. I would compare the state of the men in a world conducted by inexorable fate, with the condition of men living in a world governed by a righteous being, living under impressions of the righteous administration of all things. And I apprehend I need not say much to shew that the former is more uncomfortable than the latter. The case of the world without a wise governor is like a ship without a pilot or a compass or any hand on board who knew anything about

---

[10] Pierre Bayle (1647–1706), celebrated French man of letters and religious skeptic, resident in Rotterdam; author of widely read *Historical and Critical Dictionary* (1697), among other controversial works.

ships or sailing; the winds and tides and currents drive her hither and thither and she can pursue no determinate or regular voyage. So on the system of the Atheist, Necessity, or Fate, or Chance drive on everything in the like blind way without either intelligence or design.

Now would we not justly consider the man as distracted who would choose to make a voyage in the former, rather than under the command of an experienced master, and is not he equally mad who can suppose that it would be better for men, that all events were directed by chance than by an all wise Governor? Religion teaches us to consider the Supreme Being as the kind father of the universe, who knows our frame and pities us as a father doth his children. It is as bad then to wish to be from under his indulgent care, as if children should wish to be orphans. But every rational man is so far from this wish that he considers the existence of deity as as necessary to his wellbeing as the Sun to the Planetary System. He rejoices in a belief which is the Life of his Soul and the spring of all his joys.

Men may be divided into two classes—the *Thinking* and the *Unthinking*. Let us consider what influence the belief of Atheism would have on each of these.

As to [the] thinking part of mankind, if they are seduced into a belief of atheism, it would tend only to plunge them into distress, anxiety, and despair. He would see himself liable to many evils which he could neither prevent nor remedy. He would see himself compassed with infirmities which he could not remove—obnoxious to many dangers he could not provide against—thus would misery present itself to him on every side and after all often would some secret impressions of a Supreme Being, who would yet call him to account, come across his mind and enhance all his griefs. I do not deny that when in high spirits and hurried away by the pleasing gales of prosperity, he may banish remorse and all foreboding of futurity, but yet in his more serious moments, when brought down by calamities to which all are liable, and especially when he has a near prospect of his dissolution, then all his thoughts let loose upon him and he is plunged into despair. If he could still retain his atheistical principles, how would he rejoice in the thoughts of annihilation? But now he cannot enjoy even the comfort of this assurance.

As to the Unthinking, again, the only effect it can have upon such is to take away all restraint, to render him bolder in vice and callous to every manly feeling. All wise Legislators therefore have thought it proper to call in the idea of future justice as an adminicle to civil government. There is no example of any government where care has not been taken of the Religion of the Subjects. All governments have thought it necessary where any important affair depends on the testimony of witnesses that the religion of an oath should interpose as a proof of the truth of the declaration. All Princes and states too always confirm their treaties and contracts by the most solemn oaths. Such a general conviction of the Necessity of religion to aid the Civil government has led some to say that it was entirely a device of the Legislator and contrived by him merely the better to confirm his own authority and procure a ready obedience to his laws. This way of proceeding is at least a tacit confession of its utility.

Having said these things of the consequences that may be drawn from atheism and having shewed: 1) That though there was no Supreme Being yet it does not follow that there is no future state; and 2) That our happiness or misery in a future state does not depend on our conduct here; and 3) That even in the present life the belief of atheism has a worse effect upon our happiness than a persuasion that all things are governed by a wise and righteous governor; I proceed now to offer some arguments for the existence of the Deity. Many of these have been given by different authors; I shall give only those of most consequence, as I apprehend it is better to offer a few of most force than to trouble you with a great number.

1) Some authors have justly argued the necessity of a first cause from this, *that everything beginning to exist must have a cause.* This principle I endeavoured to shew you before was a first principle; a principle to which all who are come to years of understanding assent, and without which we could not act with common prudence for a single hour in life; a principle which was held as undisputed till Mr. Hume[11] dared to doubt it. I had formerly occasion to consider his arguments and shall not now resume what was then said. It is taken for granted therefore that [it] is either necessarily eternal without a cause to produce it, or if it begins to exist, there must be a cause of that existence, some being able to produce it; and with regard to this being, it too must either be eternal, or if not, then it must have a cause to produce it, some other being with power able to produce it. Thus we are necessarily led to a first cause of all or to an infinite succession of beings, one producing another without a cause. The last of these is evidently absurd; for an infinity of beings without a first cause cannot possibly be, because it would be a chain every link of which would be an effect which stood in need of a cause and what is true of a part is equally true of the whole. Thus are we unavoidably led to admit the existence of some eternal being, uncaused, necessarily existing and by his power producing everything we see.

---

[11] David Hume (1711–1776), Scottish philosopher, historian, and essayist; author of *A Treatise of Human Nature* (1739) and other seminal philosophical works; notorious as a religious skeptic, his *Dialogues Concerning Natural Religion* were published posthumously in 1779.

Lecture 74: The First Cause

Summary: If anything exists, something must have existed eternally without a cause. No two things can mutually cause each other, and no chain of dependent beings can support itself. The uncaused being must possess life, power, and intelligence, for lifeless matter cannot produce living, rational beings. Atheists have attempted two evasions: that the world is eternal, or that there is an infinite succession of causes without a first. Both are untenable. The world shows finitude, dependence, and change, while an endless regress—like a hanging chain with no point of support or blind men led by none who sees—is absurd. Reason therefore demands a self-existent first cause. Beyond necessity, the world's design reveals wisdom and purpose, most clearly seen in the ordered harmony of creation. What may appear as faults in nature are due to human ignorance; deeper understanding discloses adaptation and use. The heavens declare this design—especially the fixed stars, placed at immeasurable distances yet serving navigation, marking the heavens, and enabling astronomy. Their magnitude, order, and utility confirm that the universe is the work of an intelligent and beneficent Creator.

It is demonstrable then, that, if anything at all exists now, something must have existed from all eternity without a cause. For if we suppose there were two beings, let us call the one A and the other B, then if A created B, surely A could not be created by its own creature, and consequently A must be without any cause to produce it, and the reasoning is the same if to these we add a third being or if we add three thousand or three thousand millions, still the same conclusion will hold. For there is no principle more evident than this, that two things cannot mutually be the cause of each other.

Further, the same reasoning leads us to consider that which was uncaused, eternal and the cause of all other things, as possessed of Life, Power, and Intelligence. It is impossible that that which in itself hath neither life nor power nor intelligence should yet bestow them upon other beings. The same light of reason that convinces us that there can be no existence without a cause convinces us that every cause is not able to produce every effect. It is as shocking to Common Sense to say that mere inanimate senseless matter could confer sense and reason upon intelligent rational creatures as to say that things may begin to exist without a cause of that existence. I know only two ways which Atheists have taken to elude the force of this argument. The first is, by maintaining that the world as it now is existed from all eternity without any cause to produce it, or second, by saying that there has been an eternal succession of effects and causes without any first cause.

With regard to the first, it has commonly been said that it was maintained by Aristotle;[1] however, there are some doubts concerning this and his interpreters have followed different opinions, but we see that even those antients who reasoned best on the Atheistic system gave up this point. Epicurus[2] and

---

[1] Aristotle of Stagira (384–322 BC), pupil of Plato (see note 2 of Lecture 83 below) and founder of the school known as "Peripateticism"; author of the most extensive and influential *corpus* of ancient philosophical works, encompassing logic, metaphysics, physics, biology, philosophy of mind, ethics, political theory, aesthetics, and much else besides.

[2] See note 8 of Lecture 73 above.

Democritus[3] though they would not admit that the world was made by power and intelligence, yet they acknowledged that it was not eternal. And we find Lucretius[4] arguing strenuously against its eternity. Why, says he, if the world is eternal, why have we no monuments of anything farther back than a few thousand years? Can we suppose that everything beyond that short period, short indeed when compared to eternity, could have perished without leaving any vestige behind?

Though it is common to all nations to carry back their history to fabulous ages, through a pride of being reckoned the most antient, yet it is certain that we have no records that can pretend to any evidence that reach farther back than the Sacred Scriptures. The Chinese monuments, the best attested beyond comparison of all others, carry back their account of the settling of their nation by Fohi[5] till near the time of the deluge only; what goes beyond that is mere uncertainty without any evidence that can satisfy a reasonable man. Besides, it is evident in the things that we see that they are finite, dependent and changeable; one generation passeth away and another cometh. These things are sufficient to shew that the world is not eternal unless we suppose an eternal succession of effects and causes without a first cause, which was the second subterfuge of the Atheists I mentioned.

But this evidently appears to be a great absurdity. The absurdity of it hath been illustrated by several authors in

---

[3] Democritus of Abdera (c. 460–c. 370 BC), with Leucippus of Miletus (see n. 7 above), founder of the school known as "Atomism," the first relentlessly reductionistic and materialistic metaphysical system.

[4] Titus Lucretius Carus (c. 99–c. 55 BC), Roman poet; author of the influential didactic poem *De Rerum Natura* [On the Nature of Things], based on Epicurus's philosophy (see note 8 of Lecture 73 above).

[5] Fu Xi [Fu Hsi], mythical founder of Chinese civilization and writing system, said to have lived c. 2000 BC; following the sixteenth-century Jesuit missions to China, Fu Xi became widely known to European scholars under the name "Fuhi" or "Fohi."

different ways. Thus, we may suppose a chain hanging down from heaven, composed of many links, the first of which we see but lose sight of the last. Now were the question put, how is this chain supported? Would any man say, that the first was supported by the second, the second by the third and so on without end and yet that the whole was supported by nothing? Is not this absurd? It is as absurd, then, to suppose one thing produced by another and that by another and so on, because they all, taken together, make one great dependent whole and yet there is nothing left to create it.

Another way of illustrating it is by supposing a file of blind men pass along, the last of whom had his hand on the shoulder of the one next [to] him, and his again upon the third and so on till we lost sight of them. Now were it asked, who leads them? All that we see are blind yet they keep the road distinctly and go on in a determinate path, and we would conclude from this that some person who sees leads the whole, but if anyone will say, that everyone leads another without the aid of direction of any seeing person, would we not see that the position was ridiculous and absurd? As well might we suppose that blindness multiplied a thousand times would make sight, that dependence multiplied would make independence, or a cipher a real number. This argument then from the present existence of things to an eternal cause of all existence seems to be grounded on the plainest principles of reasoning and there is no exception made to it which can bear examination.

The other topic from which [we] proposed to argue the existence of a first cause or of a Deity was, from the appearance of wisdom and design which we see in the creation and in the structure of the Universe, to infer that they were at first produced and still are governed by a wise and intelligent cause. This is the argument, which of all others, makes the deepest impression on thinking men, and indeed it has this peculiar

advantage that the more we learn by philosophical investigation, it thereby gathers strength, and every new discovery,
discovers new evidences of the most excellent contrivance in the
construction of things. When we are ignorant, we may often
imagine that we perceive faults in the construction and management of things, but this is the effect only of our ignorance.
When we attend to them more narrowly and perceive their
uses more clearly, what were thought to be faults appear to be
excellences.

Some have conceived that the world would be more beautiful if there were no mountains, if all were verdant fields and
flowery meads, and if there were no rivers or seas, but this
is ridiculous; without mountains there would be no springs,
without seas, there could be no communication with distant
countries or kingdoms, and we in consequence would always
remain savages. And the same is the case with respect to all
others; what we think faults in the constitution of the world
is plainly an evidence of our own ignorance. As therefore the
more we know the more we discover marks of wise contrivance
in the formation of the Universe; this shews the importance of
improvement in the knowledge of Nature, as we thereby bring
more clearly to light the great author of all. This is an argument
which leads us to a very wide field, because every object we can
contemplate exhibits to us marks of wisdom, and as it generally
makes a deep impression on men's minds I shall dwell upon [it]
a little. I shall however mention only some of the most obvious.
I shall begin with those that are most distant from us.

The most distant objects which fall under our view are the
Fixed Stars. Their distance is such that our imagination can
hardly grasp it, yet we find that they are not only ornamental
but useful to us who inhabit the globe of this Earth. That
we may form some conception of these bodies which we call
Fixed Stars, it is proper to take notice of some principles of

astronomy. In reality philosophers have never yet accurately determined the distance of the fixed stars; all that they have done is to determine that it is not below such a distance, but how much more they do not know. The way in which they determine it is this; they can determine the distance of the Sun and by comparing the distance of the Sun with that of the Fixed Stars they can thus form a kind of conjecture how far distant they are. And till of late, even the distance of the Sun was not determined accurately, when by observing two transits of Venus, the one of which was in [17]61, the other in [17]69, they determined with certainty that the parallax of the Sun—i.e., the appearance of the Earth's semi-diameter at the Sun—was not above 8½" nor below 7½".[6] This was made sufficiently accurate by innumerable observations of these two transits. From this by calculation on geometrical principles it was ascertained, that it must be 96 or [100,000] of miles. An amazing distance which imagination is unable to grasp!

But there is another means of proving this by the velocity of the rays of light, which pass to all objects on this Earth in a time imperceptible to us; yet as all motion must be progressive it must take up some time and it has been proved to take between 7 and 8 minutes in its passage from the Sun to the Earth. But though the Sun is placed at such a vast distance yet are the fixed Stars inconceivably farther. Such is their distance that viewed from them the semi-diameter of the Earth's orbit will seem only a point of a few ", so that they are distant from us millions of millions of miles. Nor are we able to conclude that they are all at one and the same distance, or what is their respective distances, for though they all appear to us placed in the surface of the same Concave Sphere, yet this is owing to their being placed beyond the limits of distant vision, as we

---

[6] "Arcseconds," in modern terminology.

are not able to determine distances beyond a certain extent. It has been judged probable that those that appear least are only the more distant and not less, and that those of the brightest appearance and greatest apparent magnitude have this only from being nearer to us; that they may be placed at various distances through the immense void and are particular Suns illuminating other planetary worlds. But those which have the largest appearance and the most luminous are still millions of millions of miles distant from us.

How unbounded the dominion of the Universal King who made and governs them all! But how amazed are we to find that these bodies placed at such a prodigious distance are yet useful to us; they are made perceivable to us by the little organ of the Eye by means of the rays of light which move with such velocity and are so minute that several of them fall within the pupil of our Eye and are there refracted so as to form an image of the Fixed Stars. This amazing velocity of light whether considered in the motion or the minuteness of its particles is no less wonderful than the immense distances of the bodies seen by means of it. But we may [conclude] that the Fixed Stars are far from being useless; they afford us light to travel both by sea and land, by them the Heavens are marked out as it were by fixed points; without them navigation never would have been learned, without them Astronomy never would have made any considerable progress. Hence the use of the Fixed Stars to us does not appear to be casual. They were made not for the sole purpose of glimmering faintly in a serene sky upon this Earth, but they exhibit marks of wisdom and design intending them for the most beneficial purposes.

Lecture 75: Design in the Solar System

Summary: The structure of the solar system displays precise order and intelligent design. The Sun, placed at the center, illuminates and governs the planets, each moving in regular, mathematically defined orbits. Kepler discovered their harmonic motions, and Newton demonstrated that these arise from the universal law of gravitation, which decreases inversely with the square of distance and, together with inertia, accounts for elliptical orbits and predictable revolutions. The consistent motions of planets, comets, and moons follow exact rules, enabling accurate long-term predictions—something impossible by chance. Like a watch ordered by a craftsman, the heavens reveal purpose and intelligence. The Earth likewise shows adaptation to natural laws: its oblate spheroid shape balances centrifugal forces; its atmosphere sustains life, produces rain, and supports rivers; its seas enable navigation and commerce, fostering civilization. Mountains, though rugged, are essential for springs, rivers, minerals, and diverse life. Such provisions show that apparent imperfections serve necessary ends. Even in the inanimate world, wisdom and contrivance abound, and in the organic realm—particularly in plants, whose structured organization sustains growth—the marks of design become still more evident.

But in our own planetary system we perceive still clearer marks of wisdom and design. It is evident that the Sun being placed in the Centre was intended to illuminate all the other planets that move around him in different times indeed, but observing the most regular order. The antient Pythagoreans[1] talked much of the harmony of the Spheres, but had they known what is known now of the harmonic motions of the Planets they would have had much better grounds to talk on. It appears that their motions are regulated according to the strictest mathematical rules which produces a very great regularity, notwithstanding the law of gravity, by which they act one upon another, as well as are acted on by the Sun. We know not, for we have no means to know whether the power of gravity by which the Sun retains all the Planets in their orbits extends as far as the Fixed Stars; perhaps they are placed beyond the sphere of their power, or perhaps they are placed at such a great distance that the effects of it are so much diminished that they will not be considerable while the world lasts.

But though we be ignorant of its power over the Fixed Stars; yet we know that it extends to very great distances. It extends not only to the Earth but to Saturn and not only to Saturn, but likewise to all the comets belonging to this system, all of which, both planets and comets, perform their revolutions in certain periods and in regular orbits of an elliptical kind according to certain rules which are common to all which were discovered by the sagacious Kepler[2] before the reason of them was known. He discovered that they describe equal areas in equal times, and that their areas were proportional to their periodic times, that they all moved in ellipses of which the Sun was in one of their Foci, and he conjectured too that the square of the

---

[1] Followers of Pythagoras of Samos (c. 570–c. 475 BC), Greek philosopher, who taught that reality is essentially mathematical.

[2] Johannes Kepler (1571–1630), German astronomer; discovered the three laws of planetary motion named after him; helped to spread the heliocentric hypothesis.

periodic times was in the same proportion as the cubes of the distances. This has since been found true by observation and it is wonderful how happy that philosopher was in guessing at properties of Nature but still the reason of all was unknown to him.

The discovery of this was reserved for the great Newton.[3] He observed that all bodies on this Earth gravitated to it, he observed that this power also not only reached to the tops of the highest mountains but even to the clouds by which they were prevented from [escaping] entirely from the Earth. Now says he, why may not gravitation reach to the Moon; if it does to the Moon it may be that which retains her in her orbit. The general laws of motion had been discovered before his time. It had been discovered that all bodies remain in a state of Motion or of rest till they are disturbed by some impelling force, that change of Motion is proportional to the force and to the direction of the force impressed, and also that there is a reaction contrary and equal to the impelling power. These laws had been discovered before but never applied to the power of gravitation but Newton found that the same Laws extend not only to the surface of the Earth but to all the Planets.

It is evident from fact that this power deceases as you remove farther from the Earth. This has been found by experiment, because the motion of a pendulum is slower on the top of high mountains than at the surface of the Earth. It appears that this power decreases in regular proportion as the distance of the bodies gravitating to one another increases and that it decreases reciprocally as the square of the distances. It is

---

[3] Isaac Newton (1642–1727), English mathematician, astronomer, and physicist; discovered the three general laws of motion that bear his name, as well as the equation of universal gravitation; also carried out pioneering experiments in optics; author of *Philosophiae Naturalis Principia Mathematica,* or "*Principia*" for short [*Mathematical Principles of Natural Philosophy*] (1687).

a power which belongs to all the planets. They gravitate to the Sun and the Sun to them, the Moon gravitates to the Earth and the Earth to the Moon as appears by the tides. The Secondaries of Saturn and Jupiter also gravitate to their primaries. Thus was this beautiful system carried on by this simple law and carried on according to the exactest laws before the reasons of them were known. For Newton has demonstrated that supposing this power to take place, then the consequence would be that this together with a projectile force will make them describe elliptical curves, and that by this law they would describe equal areas in equal times and that the square of the periodic times would be in proportion to the cubes of the distances.

Thus we see the whole system regulated by exact mathematical rules. We see these producing the most accurate and constant operations. Now can we seek stronger marks of wisdom and design in the Universe than this? If a man sees the structure of a Watch and sees that the whole is moved by one great spring; if he sees how such wheels move such pinions, which again move other wheels and if he finds that all of these are regulated by the balance, would any man that saw this pretend to say that it was the effect of chance and produced without any skillful agent? But what is this to the planetary system in which wisdom and design so clearly appear? The Planets of our system seem to have obtained their name from their *wandering* appearance in their conjunctions, oppositions, elongations, progressions, and retrogradations. In all these cases they thought the bodies of our system wandered while the Fixed Stars seemed to remain in one plane but now men see all these wanderings reduced to accurate rules and now philosophers with the greatest ease can predict for hundreds of years their precise plane and various motions.

This was particularly seen in the last transits of Venus over the Sun when she appeared like a spot upon the Sun's disk. This had been predicted by Kepler long before, but the tables at that time were so inaccurate that even Kepler himself began to doubt that it would not happen, yet an Englishman who had attended to the Subject much was satisfied that it would happen and had the happiness to see it; Jeremy Horrocks[4] an Englishman in 1639 was the first of Adam's race who ever observed this phenomenon. We are sure that it had not been observed before, because telescopes were not in use and without them it can't be seen. However, his observations upon it were so accurate that this transit can now be foretold and has been seen since. I mention this only to shew by how exact rules all the planets are regulated; if they were not, it would be impossible to predict their appearances, but these appearances are found in fact and by experience to be regulated by unvarying rules. Now chance acts by no rules; nothing regular is produced by chance.

The learned Archbishop Tillotson[5] hath well observed, as, by carelessly throwing in a heap an infinite number of types[6] it is not to be expected that a fine polished poem would be made or even a tolerably sensible discourse in prose, how much less can we suppose that this beautiful system of Nature could be the work of blind chance acting by no fixed rules? Nor does this power of Gravitation account only for the comets and Planets in general keeping their orbits around the Sun but

---

[4] Jeremiah Horrocks (1618–1641), English astronomer; predicted and observed the 1639 transit of Venus across the Sun.

[5] John Tillotson (1630–1694), Archbishop of Canterbury; English author of numerous sermons and philosophical works; Foster suggests Reid may have in mind his Sermon CXXXVI, "The Wisdom of God in the Creation of the World," published in volume 6 of *The Works of Dr. John Tillotson, Late Archbishop of Canterbury, in Ten Volumes* (London, 1820).

[6] The image here is of individual letters set up in movable type for a printing press.

also for the irregularities in the Moon's motion, for which the antient philosophers were obliged to invent so many cycles and epicycles and which, after all, they could never fully explain. But by the gravitation of the Earth to the Moon [we account] for it easily and fully.

We know not the number of comets belonging to our system but we know that they are regulated by certain laws. These are so well known that some of their appearances have been foretold and have come according to that prediction, [though] it is difficult to determine it with accuracy till astronomy has been [accelerated] for a long series of years. One of them takes 75 years in its course and it observes its time exactly. Nothing then surely can afford stronger marks of wisdom and contrivance than our Planetary System. And if we descend to the Earth we will perceive still stronger marks of design there. The figure of the Earth as [is] best for various reasons is nearly spherical, not altogether so; the parts at the Equator are higher than at the *poles*, and the figure becomes what is called an *oblate spheroid*. It commonly was thought to be an exact Sphere, but were this the case then all the parts towards the Equator would be overflowed with Sea, the centrifugal force bringing [it] to the Equator from the poles and leaving them dry.

The wisdom of Nature appears here then in giving the Earth a figure corresponding to the nature of this power, by which we see that these parts have all their just proportion of Sea and Land and so it is over all the Globe. We find too an atmosphere surrounding the earth and which extends to 40 or 50 miles height above the surface of the Earth. This atmosphere was not made in vain. It is necessary both to animals and vegetables. Without breathing animals could not exist, even those that do not appear to have lungs; yet all find air necessary; even fish could not live without air. And it appears equally necessary to vegetables as to animals and, being thus necessary to all, the wisdom of Nature hath produced it in sufficient quantity to answer all these purposes. It invests us around like our garments and we can go nowhere when we have it not. Some antient atheists have argued that it was useless, but without it we could have no rain; the vapours are carried up and supported by it, till they fall down by the action of gravity; and without it we would have no rivers, no springs. Thus we see all the constitution of Nature admirably fitted for the care of the various inhabitants of this Globe.

Some antients also found fault too that there was so much sea; why was it not all fruitful fields, and thus supporting many more inhabitants? But in this we see the folly of Men when they begin to censure the works of God. We see it is necessary not only for furnishing supplies to rivers but also for Navigation. The wisdom of Nature intended us as Social creatures, and our Society was to be not only with those that are near but the most distant parts of the World and she therefore has furnished us with navigable rivers and with seas by which we may visit distant regions and convey their improvements and productions to our own country.

No man can trace the origin of Navigation; where seas are there, nations have been found engaged in Commerce. And Commerce is one of the greatest means of improvement among Men. Hence as far back as we can trace antient history we find those the most improved who were soonest engaged in commerce. Thus all who lived on the banks of navigable rivers or of seas were always first civilized, first improved in arts and sciences, whereas those who inhabit the heart or inland parts of a Country are long rude, thus in the heart of Asia and Africa there has been no improvement for thousands of years. On [the] other hand we see that the Egyptians who lived on the banks of the Mediterranean, the Red Sea and the Nile were early a commercial people and early flourished in the arts and

Sciences. So too the Arabians on the Red Sea; we see too the
Chinese have many navigable rivers and are a polished people.
Surely then it is not in vain that Nature has given such a great
proportion of water.

Other antient atheists have thought too that mountains
were a useless deformity on the face of the Globe. Why such
rugged rocks and horrible precipices, the dens often of wild
beasts? Would it not be better and more beautiful were it
all a verdant plain? Here too we see the weakness of man in
pretending to censure the works of God. Were the Earth a
plain then there could be no rivers, because Rivers run only
where there is a descent. All springs fall upon the laps of hills,
where moistening the Earth they descend till they come to
some strata which they cannot penetrate; they run along them
and bursting out, come down in copious streams. And without
rivers to moisten it how uncomfortable an habitation would the
Earth make!

Besides, it is evident that the mountains are fitted to
maintain some animals and so too vegetables of certain kinds
grow only on mountains. They contain also those metals which
are so useful to Man and various substances from which human
industry has reaped great advantages. Some poets indeed have
introduced metals, especially shining gold as the cause of many
evils to Man. However this may figure in Poetry, there is no
Truth in it, no solidity in it. No doubt there have [been] many
evils arisen from the use of it, but this is the abuse of the
creations of God; they tend to promote our advantage, but the
best things may be abused. And the placing of these minerals
in mountains is a strong mark of design and wisdom, as they
are thus prevented from encumbering the face of the Earth.
In great houses there are always cellars to hold what is not
immediately useful, so these may be called Cellars in which to
deposit the metals and minerals so useful to Man.

Now surely no man can call all this the effect of chance or say that there is not wisdom and contrivance in it. Thus then we see evident marks of design in the inanimate part of the Creation, that part of the Creation which is unorganized and hath neither animal nor vegetable life; but if we attend to the vegetable we find marks of wisdom still more striking. Vegetables differ from unorganized bodies in various respects. In vegetables there is some kind of organization to be seen and all the parts of a plant have a certain relation to the whole which is not discoverable in the minerals and metals. Thus a stone may be broken or divided [ever] so much but still each of these divisions is a stone. But with plants it is otherwise. They have a certain unity by destroying which you destroy the plant. We do not know in what Vegetation consists. We know indeed that every plant is filled with tubes to carry up sap and that by this means the tree gradually extends itself but how this effect is produced, it surpasses the power of Philosophy to explain as of yet.

Lecture 76: Design in Plants and Animals

Summary: Vegetable life exhibits organized structure and purposive functions: roots seek soil and moisture while stems orient to light; many plants display "sleep" movements; dormant seeds can remain viable for years and germinate with proper heat and moisture. Botanical diversity is stable across genera and species, fitted to climates, soils, and seasons, and provisioned for human and animal needs (food, medicine, clothing). Reproduction and seed dispersal show varied, adaptive mechanisms; plant organs (roots, leaves, bark, vessels) serve identifiable ends, as revealed by Malpighi and Grew's anatomical studies. Climbing plants employ specialized claspers or adhesives, and seeds are engineered to penetrate protective coats—features collectively evidencing design, not chance. In animals, perception and instinct guide species-typical behaviors (reproduction, nest-building, care of young) that preserve kinds over time; brutes lack abstraction, moral reasoning, and self-government. Anatomical adaptations match ecological demands—compound eyes, nictitating membranes, protected eyes in moles—while sensory acuities aid survival and diet selection. Many animals serve human purposes (ox, horse, dog, elephant), with dogs especially disposed to companionship. Across vegetable and animal realms, stability of species, functional organization, and goal-directed instincts indicate wise contrivance in nature.

The structure and organization of vegetables is indeed wonderful. They bend their tender fibrous roots into the Earth and creep along in quest of support, while the branches invariably ascend to the light and flourish in the open air. Besides it has been discovered by some late botanists, that many of the plants have what is called a *Sleep*, that is, at certain times they shut themselves up as it were to rest and at others they open their leaves wide to receive the influences of the Sun and the dew. They are capable of an irritation by heat and moisture, it appears, by which they acquire life whereas they are dead in the seeds. These seeds it is found can lie without life for a very long time, sometimes for years and even for Centuries; yet when they meet with a proper degree of moisture and heat their powers disclose and they burst forth in all their parts.

We know not what produces this organization which distinguishes them from inanimate matter and bodies where the life is gone, far less do we know how they grow and how they propagate their kind; but we see that they are admirably fitted for all these ends and that they are carried on by regular laws. The variety of them is great but we evidently see the intention of Nature in giving that variety corresponding to the variety of climate, soil, heat and moisture. Some we see affect the mountains, some the valleys, some one season and some another, one hot another cold, and by this means is the face of the Earth always covered with verdure and no soil is to be found without it; even the rugged rocks produce a great variety of beautiful and useful mosses. It appears to have been the intention of Nature in this variety to satisfy the uses and accommodation of Men and the inferior animals. There is no Science which has been cultivated with more assiduity in modern times than Botany, yet the Botanist has not been able to ascertain exactly the number of the various genera of plants, though to all probability, they have been fixed from the beginning of the World.

We find that this branch of knowledge was attended to by the antients and so was carried to considerable degrees of improvement by the labours of [Aristotle],[1] though in the days of Hippocrates[2] it was in a low state. But as their descriptions are so inaccurate as hardly to be understood, we are often at a loss to know the plants by the description they have given of [them]. Indeed the only descriptions that can last and be understood by posterity are those that are founded on some systematic and accurate division and arrangement of the whole genera and species, a plan which they never adopted. This however has been adopted in modern times and the descriptions now given are such as must be understood in all future ages. Future ages will thus too be able to decide with more certainty whether the same genera and species have always continued or if any have been lost. But as far as we can learn, it cannot be shewn that any one of them hath perished or that any new one has been produced.

All of us seem to have been endowed with particular qualities by the great Creator of producing each after his kind, but none have a power of producing new kinds. It is supposed that in all, there may be about twelve or thirteen thousand genera of plants added to our former stock on this subject by the late discoveries in the South Sea. All of these have distinct characters by which they may be distinguished from any others and may be described so as to be known by the description. Now surely all this cannot be the Effect of Chance. They all observe established rules by which it appears that an all-wise cause first formed and still carries them on in their operations.

---

[1] See note 1 of Lecture 74 above.

[2] Hippocrates of Cos (c. 460–c. 370 BC), the "Father of Medicine"; the writings of the Hippocratic *corpus*, including the Hippocratic Oath, were conventionally attributed to him.

We see some of them fitted for food to Man and the other animals, others for Medicine, others for Clothing.

There is no use of human life [that] can be thought of for which some of them are not properly fitted. And even those plants which are poison to some are wholesome food to others. Further—as they are the common aliment of all, both men and brutes, for all either live on vegetables or on some animals that are supported by vegetables—it was proper that there should be such a quantity as is sufficient to support the life of all the inhabitants of the Earth, and this accordingly nature has done. Some of them require no care of ours in order to rear them, others again require culture to bring them to perfection, and it evidently appears to have been an employment intended by Nature for Man, that he should cultivate the various plants of the Earth, discover their qualities and fit them for his use; the first appearance of this culture was in the Garden of Eden and it still continues an important employment to supply us with the necessary supports of human life.

Further they are all fitted with powers to propagate their kind that none may perish. In the manner of producing this seed and of disseminating it we find a great variety. In some the seed is strongly guarded by an oily coat, which defends it [against] external injuries and admits moisture that is sufficient for its growth but no more. Some of them are wafted through the air by down, others are thrown at a distance by the elastic spring of the seed, some are carried away by birds, and so on. Some of them we see are of a large, some of a small size. How uncomfortable would life be without trees sufficient for houses, ships and various utensils that render life agreeable? As far back too as we can trace the history of men they appear to have used some kinds of vegetables for Clothing, as flax, hemp, etc.

It may here be observed, that there is in vegetables an unaccountable disposition, that when the seed is planted in the

Earth, whatever is its position, still the roots push downward and the [stems] upward, but by what attraction or repulsion this is produced we cannot learn. We see that in the seed there are two opposite ends from which the stem and the roots issue out, and yet though the stem is placed downwards, yet it will turn round and pushes upward; on the other hand, though the roots be up, yet will they bend downward and search into the earth. Nay, we may observe that if a root is planted amidst earths of different kinds, its fibers spread around, avoiding the bad and seeking the best. Now we have no reason to ascribe to vegetables either sensation or thought; in this they are guided by the power implanted in them that is necessary to propagate their kind and provide against the species being lost.

Neither are we able to explain the manner in which they draw their nourishment from the Earth, every one drawing that which is proper to itself. For this purpose they have a wonderfully curious structure. Their roots are divided into very small fibers which receive the sap from the ground, which is conveyed in sap vessels, and is afterwards curiously altered and assimilated to the Nature of the plant by vessels of various kinds. This has been fully and accurately observed in modern times by Malpighi[3] and Dr. Grew.[4] These two gentlemen formed the design of examining the structure of vegetables about the same [time]. Being both members of the Royal Society they transmitted their discoveries to them by whom they were published, and though they carried on their researches separately, yet their descriptions so remarkably agree as to add the greater authority to both. The engravings of these two philosophers

---

[3] Marcello Malpighi (1628–1694), Italian anatomist and microscopist; the first to see and describe numerous microscopic structures in plants, animals, and embryonic development.

[4] Nehemiah Grew (1641–1712), English botanist and microscopist; author of the great, four-volume *Anatomy of Plants* (1682).

shew that the works of Nature are continued at once in a beautiful and useful manner.

It may be observed that particular parts manifest [the] intention with regard to the whole; now every intention supposes design and intelligence. Every one sees that the roots are designed to furnish the plant with nourishment and to fix it firmly in the ground. The leaves also in time of a drought draw in moisture from the air. No man can doubt that the bark was intended for a covering to the tree and it serves admirably for this purpose, and the inmost bark is the part from which the growth arises, it being a film of the inmost bark that swells from year to year and makes an addition to the body of the tree. We see some of the seeds enclosed in a shell so hard that we can penetrate it with difficulty; we would be apt to think then that it must rot in that shell, but Nature has provided against this, for always opposite to that point from which the root issues it is so soft that the root itself is able to force a passage through it. Thus we see that Nature has provided against every difficulty, for that shell which to us seemed hardly penetrable is easily perforated by the slender fibre of a seed.

Some plants too we see are not able to support themselves; they must be supported by other things, thus they cling to trees and walls, whatever can bear them. Of this kind are, ivy, vines and hops. For this peculiarity nature has provided in their structure; some of them are furnished with a kind [of] claspers which creep around the body that supports them; such are the hops and the vine. Others again exude from themselves a sort of dung which glues them to the body that bears them, as the ivy sticks to the rocks, and they adhere so firmly as to resist all the violence of the winds. It would be endless to mention all the marks of wisdom and design which appear in the vegetable creation. What I have said may suffice to shew that they cannot be attributed to chance.

I shall now consider a little the marks of wisdom to be seen in the animal creation. Here I shall divide what I have to say into three parts, 1) The marks of wisdom and design in the structure of the lower animals; 2) The structure of the human body; 3) In the human Mind. In animals, the qualities of both inanimate matter and vegetables join but they have something of a superior Nature to vegetation still. All of them have some perception of external objects and some of them too have a degree of Memory or something very like it; thus a horse will know the way home again and will keep [to it] even though it is so dark as that the rider himself should not know it. They appear too to have some trains of thought though we cannot discover their laws. We see among animals a great variety in different cases, each keeping the way prescribed to it by Nature. Some are viviparous and suckle their young; some again are oviparous and hatch their young by incubation; yet all agree in breeding them carefully after they are produced.

By this Nature takes care that while the individuals are always perishing yet that the species should not entirely perish; accordingly in all the variety we know of quadrupeds, of birds, of fishes, insects and rep- tiles, it cannot be shewn that any one kind has perished altogether. This however will be more certain in future ages, because by the industry of the moderns, they have been more accurately reduced to genera and have been most distinctly described by those who have applied to this branch of knowledge. It were to be wished that Natural historians were more care- ful in describing their various instincts. They have been laborious in describing their structure, but surely the instincts by which they live, and by which they are preserved and regulated are no less worthy of observation. These have been more or less observed in every age.

Aristotle has treated this subject; he made his collections and observations with candour and judgement, being furnished with expenses by his pupil Alexander.[5] His descriptions are faithful where he himself made the observations; where he did not, he delivers it only as a hearsay or report. The descriptions he has given of what he saw are just and answer exactly at this day. So true it is, that though man may in several respects [be] said to be not the same as in his days, yet in brutes there is no difference; they are the same now as in the days of Julius Caesar.[6] They are not able to communicate their knowledge and for this cause never can arrive at higher degrees of perfection, but by instinct they are fitted to preserve themselves and to continue the Species. And though some species are made the prey of others yet we do not see that they ever perish, Nature having made them more prolific in proportion to the devastation made upon them.

There is no evidence that the brutes possess some of those powers which distinguish human creatures [such as] Abstraction, moral perception, and reasoning, nor do they appear to have any power of Self-government; they seem to be always directed by what gives the strongest present impulse, nor do they appear to pursue any rules of action to the attainment of any end. It is true we see a kind of government among black cattle, but they appear to be governed by instinct rather than by fixed laws.

And how prodigious so ever the variety of animals may be, yet have they always continued from age to age perfectly

---

[5] Alexander III of Macedon (356–323 BC), known as "Alexander the Great"; taught by Aristotle (see note 1 of Lecture 74 above), he established an empire encompassing the entire region from Macedonia in the north to Egypt in the south, and from Greece in the west to the Indus River (in what is now Pakistan) in the east.

[6] Gaius Julius Caesar (100–44 BC), celebrated Roman military leader and politician; after conquering Gaul (France) and Britain, he seized power in Rome through a military coup, leading to civil war and ultimately the end of the Roman Republic.

regular in their way of life, every species its own way. We may here take notice only of the way in which birds build their nests. How great a variety is there in their way of building, yet do all those of a species build in one way. They all build in a place which appears free from danger or disturbance, where they may quietly bring forth and safely rear their young. In this last, both the parents commonly join but in some cases where the care of one of them is sufficient the other leaves off attending to them; and even there are some cases where the young need no aid and there the parents leave them to shift for themselves. Thus the caterpillar.

From all this we see what an infinite number of instincts belong to the brute creation and all precisely suited to their manner of life with great skill, and of this we can have no solution, but in concluding that they were so ordered by the wisdom of him who ruleth over all Nature. If again we consider their bodies we will find a vast variety, but at the same time the structure of each admirably suited to their way of life.

Thus, we see that some animals such as quadrupeds, who can easily turn their head, have their eyes placed in the side of the head, but others who have not this power of turning their head, have eyes all round, and behind as well as before. Some flies and bees too have two crusts like hemispheres on the sides of their head, in each of which are inserted a great number of eyes; in each of them there is a distinct picture formed, each of which have a distinct optic nerve and thus they see in every direction. In Spiders some have 4 eyes and some have 6. In the Mole they are all covered with hair so as to be safe in pushing through the earth and are so small as to be hardly discernible yet are sufficient to warn them when they are above ground. And in those animals that have occasion to push among trees and brushwood the eye is defended with what is called a *membrana nictitans*.[7]

In the other senses too there is the same beautiful variety and wise contrivance. One being endowed, where their situations require it, with a wonderful acuteness of Sight, another of Taste and so on, by which they may distinguish wholesome from poisonous food. For this reason, depending on *their* taste, when people are shipwrecked in warm climates and come to woods where varieties of fruit present themselves, they think they may safely eat those in which they see marks of the birds beginning to feed upon, but those that are untouched they suspect as noxious. It is evident that some of the animals are intended for the use of others, to serve them as prey, but still more were intended for the use of Man. The ox was intended to assist him in his laborious employments, the horse, the dog and the Elephant were all designed for the service of Man and they have generally such instincts as fit them for this purpose. The dog especially seems intended to be his companion. If we can trust to Colben,[8] a traveller, there are great numbers of dogs that run wild in troops without a master and though they prey on every kind of beast yet they never prey upon man; nay, so great is their reverence for him that they will allow him to carry off what prey they have caught without hurting him.

---

[7] "Nictitating membrane"; a transparent or translucent membrane that can be drawn across the surface of the eye, providing protection while still admitting light; present in many birds, reptiles, and mammals.

[8] Peter Kolbe (1675–1726), Dutch explorer; author of *Caput Bonae Spei Hodiernum* [*Today's Cape of Good Hope*] (1719).

Lecture 77: Design in the Human Body

Summary: Reid argues that human anatomy exhibits fixed laws, proportioned means to intelligible ends, and therefore marks of divine wisdom rather than chance. The alimentary system shows purposive structure: the mouth (tongue, teeth, salivary glands) prepares food; infant dentition develops gradually and teeth differ by function (incisors, molars); swallowing proceeds without art; the stomach, bile, and lacteals effect digestion and absorption. The circulatory system—discovered as such by Harvey—moves blood from veins to heart to lungs and back to arteries; valves prevent reflux, and arterial–venous communications maintain flow after injury. The nervous system arises from brain and spinal cord, well protected by skull, dura and pia; pervasive nerves mediate sensation and motor power, though their mode of action is unknown. Muscles contract under neural influence and are mechanically proportioned to their tasks (Borelli), lubricated by membranes and fat; the skin and cuticle protect the body. The senses are fitted to their objects: the eye forms an image on the retina; the ear conducts vibrations via tympanum and ossicles; olfaction and taste are distributed over nose, tongue, and palate. The skeleton develops from softness to strength, with joints and sutures adapted to varied motions and stability. Such systematic contrivance, Reid concludes—echoing Scripture—implies an intelligent maker, a view he notes even Galen came to affirm.

But it would be impossible to enumerate every mark of wisdom to be met with in the animal creation for every part shews it. These I have mentioned seem to be regulated by fixed laws; they manifest certain ends to which there are proportioned means, and must satisfy every candid mind that he who made them is divine, and that chance never could produce them. I shall now consider the marks of wisdom to be met with in the Body of Man.

It is impossible not to see that man was intended to take care of his own preservation, by food and drink and by alternate labour and repose and we see that his constitution is fitted for this purpose. As 'tis necessary that his body should be supplied with food, this is taken into the Mouth which we find adapted to this end, to prepare it for digestion in the Stomach. The mouth is furnished with the tongue, teeth, and glands, all of them admirably contrived for their several ends; the teeth in order to grind and comminute it, while mixing with a fluid secreted by the glands, it is fitted to pass through the gullet without [injuring] it. We see likewise that these teeth are not perfectly formed in infants; this would be inconvenient for the Mothers in suckling them; they are only a pellucid mucilage in that socket or cavity in the jawbone which afterwards receives them and holds them so firmly. By degrees, they grow into a greater consistency as [a] small opaque speck appears in the middle and they gradually harden and push up till they penetrate the gum.

It may be remarked that all our upper bones are covered with a covering called the *periosteum* which is necessary to nourish them, but the teeth have it not; for though they had it, it would soon be worn down and destroyed and so be of no use. They appear of different forms and are designed for different purposes, some to cut and others to ground the food we eat; hence they are divided into two kinds by the Anatomists, the *incisores* and *molares*, each of which we see are fitted for the ends intended. And that man surely must either be stupidly blind or obstinately perverse who does not see that all the structure of the mouth is[1] the most proper of any that could be thought of, to receive our food and [prepare] it for being digested.

The preparations too made for swallowing our food are admirable. Were we left to do this by our own art we would of necessity starve, but we see it is done without any care of ours. The Stomach also is evidently intended for digesting our food, being by its structure fitted for the purpose and [having] certain glands in order to assist it. We know likewise that the bile is thrown into the Stomach and mixes with the contents of the stomach and guts so as to help the digestion. There are innumerable small vessels which are called *lacteal*, fitted to receive what enters into them while the rest is carried down and goes off. What is separated by these vessels is carried by other vessels and mixes with the blood in the left subclavian vein where by means unknown to us it is assimilated with the blood and makes that fluid which is so necessary to our existence. The system of the veins and arteries have also a most wonderful construction. This blood being collected together by the veins, and brought to the right auricle of the heart, is from thence conveyed by the pulmonary vein to the lungs—through which many branches of that vein are spread—where mixing with the air, it is fitted for our support. From thence it returns to the left auricle of the heart and from thence is thrown into a great artery.

The muscles of the heart have a strength proportional to the work they are to bear. They contract and dilate alternately with great force without any intention on our part. But as we

---

[1] Baird wrote "is not," but sense seems to require the deletion of "not."

know from the principles of philosophy that the blood exits equally in all directions, we would be apt to think that it would return back, but nature is never deficient in any of her operations; she has guarded against this by valves which afford an easy passage out to the blood but hinder it effectually from returning, so that it is still protruded forward and the arteries being constructed with valves of a similar kind, it is thus carried to the utmost extremities of the body. This circulation of the blood was unknown to the antients. We owe the discovery to the famous Dr. Harvey,[2] Physician to Charles I.[3] It is now universally adopted and is sufficiently evident in men and all the other animals that seem to have blood. In all we observe, a fluid which though not red, supplies the place of it. The construction of the vessels through which this fluid circulates is exquisite and wisely contrived. No hydraulic or Hydrostatic machine could better serve their purposes. And as we are ready to admire any contrivance for conveying water from one place to another and the like, so ought we to admire the invention of Nature in adapting the vessels of the heart, the arteries, etc. to the circulation of the blood. But from this common mass of blood there are other fluids to be secreted by the glands, the whole mystery of which we are unable to discern.

As there is a system of veins and arteries for circulating the blood which divide into smaller and smaller branches till they reach the farthest extremities of the body, so nature in case they should be hurt by bruises or even destroyed by amputation perhaps, has provided against these contingencies by making communications in many places between the arter-

---

[2] William Harvey (1578–1657), English physician and anatomist; discoverer of the circulation of the blood; author of *Exercitatio Anatomica de Motu Cordis et Sanguinis in Animalibus* [*Anatomical Exercise on the Motion of the Heart and Blood in Animals*] (1628).

[3] King Charles I of England (1600–1649), reigned from 1625; deposed and ultimately executed during the English Civil War (1642–1651).

ies and the veins and between the arteries themselves. The communications, if an amputation, are at first small, but they widen by degrees and the circulation goes on as freely as before —Another mark of wisdom in the structure of the Body is the system of veins and nerves. These we know are the instruments both of sensation and of muscular motion though we know not how they perform their offices. The brain from which many of them issue is admirably guarded by the bones of the head which enclose it and defend it from external injury. These are at a small distance from each other in infants and are capable of a small degree of compression, but by degrees become so indented in one another and become so firm as to form a firm covering to the brain. This is also guarded by two coats called the *dura mater* and the *pia mater* and from these two coats proceed films which go along with all the nerves that proceed from the brain.

But besides the brain there are nerves which proceed also from the *Spinal marrow*. These too are guarded by the backbone from injuries, but as the back is often obliged to be bent, it is formed that it may yield without hurting the *marrow*. From these two, the brain and spinal marrow do all the nerves proceed in pairs. It is commonly thought that there are about 39 or 40 pairs of them, for in some particular person there may [be] a small difference, yet in general they are similar, so that the descriptions given in anatomical books are commonly found to answer. The nerves extend to all parts of the body, so that a pinpoint cannot be set down without touching some of them and they are so distributed because they are the instruments of sensation, so that if a nerve is cut it has no more feeling than if it were not a part of the body. They are [thereupon] divided into smaller and smaller branches till they become imperceptible to the sight. How they perform their office, Anatomists and Physiologists never have been able to discover. They do not appear to have any fluid in them like the veins; they are rather solid, though of a soft and medullary substance. But though we know not the manner how they perform their office, yet by the effects, we see, that they are fitted for it.

From the nerves the power is conveyed to the muscles, by which they contract and dilate and so produce all the motions of the body. The muscles are of a fleshy substance, being more so in the middle and a kind of tendons round it by which they are drawn together and perform their motions. The famous Borelli,[4] an Italian Philosopher, calculated the strength necessary in our muscular exertions. Their strength indeed must be prodigiously great, because by a Law of Mechanics, the nearer a force is to the fulcrum or centre of motion, the greater must it be, and for the conveniency of the Body, and to prevent it from being too bulky, the force exerted by the muscle must be very near the centre of motion. The muscles of the arm for instance must be within an inch and a half of their centre of motion and yet the force to be raised may be two feet in distance, so that the strength of the muscle must be to the force as two feet is to an inch and a half. In this Borelli calculated the force necessary for all the muscles and we see that Nature has adapted them exactly to it. They are not insufficient for the force they require for there are no instances of them being broke. They have been shewn by anatomists to be admirably fitted for giving motion to the body.

How they are contrived is beyond our comprehension, but when a muscle acts we know that it swells in breadth and contracts in length and thus the motion is produced. We know that this power is communicated to them by the nerves, but

---

[4] Giovanni Alfonso Borelli (1608–1679), Italian mathematician, physicist, and physiologist; the first person to apply the new science of motion to the physical aspects of the operations of the human musculoskeletal system; author of *De Motu Animalium* [*On the Motion of Animals*] (1680).

of the manner of its operation we are entirely ignorant—so far is the wisdom of God beyond the wisdom of Man. That these muscles may move easily without injuring themselves or one another, they are all surrounded with a membrane which is commonly moistened with fat, that at once lubricates them, but also fills up all the interstices so as to add beauty to the whole. When this is taken away, and the muscle made bare the Body appears a most horrid spectacle. We may observe that the whole body in order to be preserved in its curious organization, must have one common cover, and here Nature has also shewn her wisdom and design in making a skin, which is a tough membrane capable of great contraction and dilation, so that whether a person is fat or lean, young or old, it affords a close cover.

This skin is preserved by a *cuticula* or scarfskin and as this is much eexposed to injuries it is renewed when rubbed off. It consists of scales which are easily perceived by a microscope. It is this cuticula that is raised by a blistering plaister, and we then see how tender [is] the real skin and how impossible it would be to live without this. But we must not omit that besides the nerves that are for the muscular motions there are others for the various senses, one pair called the optic which are the instruments of Seeing, another called the [olfactory][5] which is the instrument of Smelling, the auditory for hearing, and other small fibres of nerves which go [to] the various parts of the body and are the instruments of touch. These however are collected in the greatest number at the points of the fingers. We are at as great a loss to say how these nerves perform their functions as those that are intended for the muscles. We see no difference in their construction, and yet we see that one kind

---

[5] Baird left a blank space, but "olfactory" seems clearly to be the word intended.

are fitted for giving sensation and perception, and the other only for giving motions.

There are holes in the skull and spine exactly fitted for the transmission of the nerves to all parts of the Body. The optic nerve enters the bottom of the orbit of the Eye and is fixed in the globe of the Eye, where if it is cut or obstructed there is no distinct vision, though this eye be perfectly sound. The same happens of all the other nerves, though we are ignorant what impulse is necessary towards their communicating sensation and Perception to the Mind.

Besides the nerves for sensation, there are also external organs of admirable contrivance. The Eye is an organ admirably fitted for Vision. It is necessary that a picture of the object should be formed on the coat called *Tunica Retina*. How this conveys the image to the brain anatomists know not, but this they know, that when the image is not properly formed the vision is hurt or destroyed. It would be difficult perhaps to make those who do not understand anatomy comprehend all that is known of this little organ. I shall not therefore enter on the task but those who have the least discernment will observe that it is intended for seeing and therefore that the rays of light are fitted for it and it for them with admirable skill. Nothing can be more surprising than that by a small ball fixed in a socket we can perceive the fixed stars and the various objects around us by means of the refraction of the rays of light. These rays move with such rapidity as are adapted to this end.

The other organs of sense contain no less marks of wisdom, though we are more at a loss with regard to the use of their several parts. The external ear is well fitted to receive the undulations of the external air that produce Sound. These are conveyed to the *Membrana Timpanica*. In the inside there are several small bones that receive the impression made by the motion of it, and so it is conveyed to the auditory nerve which is spread over the inner ear. The nerve for smell is also spread over the internal part of the Nose and the nerves intended for [taste] are diffused over the tongue, palate, and circumambient parts.

Every part there is fitted for its proper use. If therefore upon seeing a curious engine we conceive that it had a wise and skillful Maker, must we not in a much higher degree apply these qualities to [the] contriver and maker of the curious fabric of the human body. This argument is elegantly summed up by the Sacred Writer. He that made the Eye, shall he not see? He that made the ear shall he not hear? He that gave a man understanding, shall he not understand?[6] No argument can be more forcible to any candid and ingenious mind.

I might also take notice of the structure of the bones, these supports of the human body, so admirably constructed for use. While the body is in the [womb],[7] the bones are soft and flexible without any of that strength and solidity which they afterwards acquire; but as they are intended for supports, they gradually acquire a firmness of texture which is in no other part of the body. They have various articulations, some of [them] resembling a ball and a socket, fit for turning in all directions as, for instance, those that join the arm to the shoulder and the thigh bone to the *os sacrum*. Of the other *joints* there are different forms, all suited to their ends, and by all of these joined together we are fitted for walking, running, jumping, stooping and the like, which we do with great ease and facility.

Besides these articulations for motion, there are other parts not intended to be moved, where the bones are not jointed, but firmly joined to one another, in a manner something like what the Carpenters call *dovetailing*. This is the case with the two parts of the Skull. In some articulations the motions are

---

[6] Cf. Psalm 94:9–10.

[7] Baird wrote "foetus," but the sense clearly calls for "uterus" or "womb."

intended to be small, such as the back, and accordingly the bones of it are fitted so as to move a very little. In some cases again we see that motion is performed merely by the bones being joined by a cartilage or intermediate substance between bones and flesh. In this manner are the ribs joined to the breast bone. We may observe too that all the bones which touch one another are smooth and lubricous at the ends so as to make their motion more easy, all of their structure exactly answering to the laws of Motion.

It is worth noticing before we conclude that the famous philosopher and Physician [Galen],[8] having been brought up on the Epicurean[9] tenets of a fortuitous concourse of atoms and that there was no providence, no care of a Deity, when he had occasion to consider the Structure of the human Body, was soon converted from this principle and convinced of the existence of a wise and designing Being.

---

[8] Galen of Pergamon (129–c. 200 AD), Greek physician, anatomist, and philosopher; author of numerous treatises that were highly influential for medical practice during Late Antiquity, the Middle Ages, and the Renaissance, in both the Christian and Islamic worlds. (Note that Baird wrote "Galileo," which in the context is clearly mistaken.)

[9] Follower of Epicurus; see note 8 of Lecture 73 above.

Lecture 78: Structure of Human Mind and Society

Summary: Nature, having provided for bodily healing, also reveals wisdom in the structure of the human mind through instincts and faculties suited to distinct ends. (1) Self-preservation: Humans are equipped with appetites of hunger, thirst, and rest, aversions to harm, and pain as a warning to avoid danger. Infants act by instinct—suckling, crying for aid—demonstrating an innate care for life. (2) Continuation of the species: Sexual affection and parental love ensure reproduction and nurture of offspring; maternal tenderness, aided by milk and infant instinct, secures survival. (3) Social life: Man is a naturally gregarious being, fitted for community by language, imitation, and social affections (gratitude, compassion, friendship). Even malevolent passions presuppose social relations. (4) Improvement in knowledge and arts: Curiosity and credulity in childhood foster learning; habit transforms difficulty into skill, enabling progress in arts and conduct. Pleasant emotions attend benevolence; uneasiness accompanies malice, encouraging virtue. (5) Political society: Differences in talent and temperament fit individuals for varied roles and governance; ambition and docility sustain order. (6) Moral progress: Conscience, self-approbation, and remorse provide internal sanctions, directing toward wisdom, justice, temperance, and fortitude. These faculties collectively display intentional design, equipping man for preservation, society, knowledge, and virtue.

To what I have already said I may now add that provision which Nature has made for the cure of diseases. As we are liable to many accidents which are apt to produce disorders in our bodies, the wisdom of Nature hath likewise provided for the cure of these. When any of our bones are broken there is a fluid which issues out. At first it is of a cartilaginous nature; by degrees however it hardens and becomes as firm as before. And all that is done by the Physician is not to disturb nature in her operations. It is the same with the cure of our other diseases. It is the operation [of Nature], not of Medicine; all that [Medicine] does is to exclude the ac- tion of the air or whatever may prevent nature in her process. This seems to have been the conception which Hippocrates and the antients had of Medicine, that it only was useful in aiding Nature without pretending to assume the merit to itself. We see too when extraneous bodies are anyhow admitted into the body, they are frequently expelled in a very wonderful manner. Of this there are innumerable instances in the history of Medicine. But I now proceed to the last branch of this division, viz, the marks of intelligence and wisdom to be found in the structure of the Human Mind.

In the structure of the human Mind we may perceive various intentions and at the same time observe means fitted to answer those intentions. And

1) It is evident that man was intended to take care of his own preservation, to avoid those hurts that would impair his health or endanger his life, and also that he should seek what is necessary to support his life. For this purpose the infant is provided with various instincts to take care of its life. It sucks when it is hungry and swallows its food, which by art it could not learn to do; it is by the instruction of Nature. When it ails too it cries by instinct, and these cries are understood by the Mother or those around, and their bowels are moved to give it

all the assistance in their power. Nature has taken care that all our diseases should be attended with acute pains which lead us to [avoid or] remedy them. As food and drink are necessary to supply the waste of the body, so the appetites of hunger and thirst are given us to fulfill this end, to admonish us when they should be gratified, and without these we could not know when, what, how much, or how often to eat or drink. As alternate labour and alternate rest are necessary to our health and even to our life, so nature has given us a disposition to alternate exercise and repose, by admonishing us when we have continued too long in either. The love of life and an aversion to whatever has a tendency to destroy it is implanted in all animals, in order to induce them to take care of their life, and so we find that there is no animal but what uses all the means in its power to preserve or prolong its life.

2) Another instance of wisdom in the structure of the Mind is an intention that the race should be continued. This we see, too, in the other animals as well as Man and they are all accordingly fitted with appetites to answer that end. In the human race the love of the sexes and the parental affection serves the end. In all ages whether men are wise or foolish, virtuous or vicious we can see no reason why the race should ever cease. In all situations Nature has provided against this and indeed if it were otherwise the race must very soon perish as individuals are only short-lived and temporary beings.

For this purpose too Nature hath furnished the Mother with Milk and has taught the infant by instinct to suck it. Whether we consider the weaknesses or the wants of children it is evidently impossible that they could support themselves without the tender care of the parents, particularly of the Mother who often deprives herself of rest and all the conveniences of life to supply its wants. And this takes place not only among wise and enlightened nations—they might perhaps do it from a principle of duty—but it takes place among all kinds of men. The different dispositions of the sexes, too, seems to have designed them for family society. This has been taken notice of in very antient times. Thus in the Oeconomics of Xenophon,[1] we find Socrates[2] display this very elegantly and agreeable. Nature has given to the female sex that timidity and delicacy that is proper for the management of domestic affairs and the rearing of the tender offspring; to the other sex, that fortitude and courage necessary to procure subsistence for the family and which require greater labour and robustness.

3) Nature evidently intended Man for Society. Solitude and seclusion appear unnatural and contrary to the constitution of his Nature. We see that all men in all ages have lived in this way, if we except a few individuals who either from affectation of singularity, or a wish to be thought remarkable for sanctity, or perhaps from false notions of Religion, have lived by themselves. But the intention is that we should live in Society and this intention is to be seen not in man only but in some of the lower animals; some of [them] we observe are gregarious, others are solitary. Foxes, lions, bears, etc. associate only during the time that is requisite for copulation and the rearing of their young; black cattle, etc., on the other hand, are naturally gregarious and are always found in Society. Now man is evidently of the last kind; he is a gregarious animal. In all ages the principles of his Nature have led them to this state.

---

[1] Xenophon of Athens (c. 431–354 BC), Greek soldier and historian; author of the famous *Anabasis* [sometimes rendered as "The March Up Country"], as well as various Socratic dialogues in the style of Plato (see note 2 of Lecture 83 below), including the *Oeconomicus*, the subject of which is household management.

[2] Socrates of Athens (c. 470–399 BC), often considered the founding father of Western philosophy; main character appearing in the dialogues of his pupils, Plato (see note 2 of Lecture 83 below) and Xenophon (see previous note); he left no writings of his own; put to death for "corrupting the youth" of Athens.

Even the most rude and barbarous are always found in tribes or clans. Accordingly we see that he is fitted for this situation.

i) By language which is peculiar to man. For though there are some signs by which the lower animals can communicate their feelings in some degree—thus, a dog can easily by his appearance warn another whether he approaches him with a friendly or a hostile design—yet these signs are few; they are all natural signs understood from their nature and it is not in their power to enlarge them. But man besides the use of natural signs, is able by them to form other artificial signs and by these communicate to others not his present feelings only but his past knowledge. And thus by receiving the experience and knowledge of others to assist his own, men are perpetually improving in civility and useful Science. And this seems to be peculiar to man only for it is not to be found among the brutes; they never improve—they are the same now as in former times and will remain the same to the end of the World. By language then it appears that we are intended for Society as it could have no existence without it.

ii) Men are led by natural instinct to imitate the actions of those around them. Now this has an evident relation to Society. For did we not live in Society we would have none to imitate. I might [pause] here to take notice of the social affections of which there could be no exercise without society, such as gratitude for favours received, compassion at the distresses of others, ambition of superiority and power, friendship, esteem, and benevolence, all of which serve as many ties to bind men together in social union. Nay, even those affections which we call malevolent shew that we were intended for Society as without it we could have no occasion to exercise them.

4) Nature intended man to improve in Knowledge and the useful arts.

i) We see that he is fitted for this by his very constitution. We observe in children a great curiosity to enlarge their knowledge; they pry into whatever is unknown, everything that is new delights them, they examine it on every side and are thus daily acquiring new ideas. Children are led to this by instinct and indeed if it were not thus they never would acquire any knowledge at all. ii) What tends to our improvement in knowledge and the arts is that credulity incident to children, by which they receive with implicit submission whatever is taught them. One would be apt to think at first sight, that we ought only to believe what we see just reason to believe when we have sufficient arguments to induce our belief. But were this the case with children the consequence would be that they would lose much of what tends to improve their faculties and enlarge their knowledge.

One of the most remarkable parts of our constitution is the power of acquiring habits, by which, by doing a thing frequently we acquire a facility of doing it. This is a power so familiar to us that we require no account of it, but if we attend to it we will find it altogether unaccountable, though we see the ends for which it is intended. Inanimate machines by going in a road frequently never learn to move more easily in their road, but man is so made, that what at first was difficult by repetition becomes easy. By this much time is spared for our acquiring other habits and arts. If children by repetition did not acquire a facility of doing them they would never walk tolerably or speak properly all their days. The same may be said of all the other arts, of writing, dancing, fencing, etc. Habit is the foundation of them all.

Further, man was intended to improve in his social affections. For this purpose Nature has annexed a pleasant sensation to the exercise of them. There is a tranquility attending them which is the comfort of life and invites us to the practice of them as our highest happiness. On the other hand, to our malevolent affections, there is an uneasy sensation annexed which admonishes us against the indulgence of them except where they are absolutely necessary.

Further, Nature has intended us for *political* Society. Though there be some very savage tribes which give small marks of this, yet it is manifest that we were designed for this. For some submission and subordination is necessary in order both to defend us against injuries we might receive from those of our own community or of a different one. Even in the rudest savages as in the *Canadian* tribes we find that though they have neither fixed laws nor magistrates, yet it is always understood, that when an injury [is done] to any of them by one of the same tribe then he is entitled to revenge the quarrel, and if from any of another tribe, then the whole tribe think themselves bound to assist him in procuring retaliation. And when they go a-hunting or to war there is always a kind of subordination, which though rude, yet is sufficient for them.

But as men improve in politeness their wants increase as stricter union becomes necessary. The indications of Nature which shew that we were intended for this state are: i) We see that men are endowed with various and different talents. Hence, we see that all are not equally fitted for every profession. This difference is not to be observed in the other animals. We see a greater difference between one man and another than between any two of them. As in a building we see stones of various figures and fit for different ends, some for cornerstones, some for lintels, some for windows and so on, so in society [various] persons are endowed with different talents and are fitted for different offices in that Society. And if they are endowed with these, Nature surely intended that they should have an opportunity to exercise them. If all lived separately and in solitude, then all would have the same things to do, so that the same abilities would be required in all, whereas in Society the deficiencies of one are supplied by another and all find a profession to suit his talents. Besides, ii) In large societies we find that there are few who have either the wish or the talents to govern. The *many* are tame and easily led by a superior address. If all had the same ambition for power and [predominance] then it would be impossible that Societies should subsist. This rivalry would effectually hinder it. But if we see that the greatest parts are easily made obedient to an ambitious few, then from this we may collect that nature intended them for government and political society.

The Last intimation of Nature, which I shall mention, is that men were designed to have the means of improvement in Virtue and moral goodness. That we have temptation to do wrong is no doubt true. It does not appear to have been the intention of Nature that we should be free from temptation. This is a state of probation; but still we have sufficient inducements to improve in virtue. For this end everyman is endowed with natural Conscience, which points out to him in most cases what is right and what is wrong. Self-approbation follows the performance of virtuous actions, but no man can perform any crimes of an atrocious nature without remorse. Every man sees that wisdom, prudence, justice, temperance and fortitude tend to real happiness, and if at any time he yields to the force of temptation and acts contrary to the direction of Conscience, he always finds remorse follows it more than sufficient to counterbalance all the pleasure he felt in deviating from his duty. Every one then has the means of improvement, though some perhaps from various circumstances, as good education, good example, etc., may have it in a higher degree than others.

From all this it is clear that in the structure of the human mind there appear various intentions and there are, we have seen, means fitted to answer these intentions, which shews wisdom and skill in the contrivance of our Constitution.

Having now at some length pointed out these marks of wisdom and design which appear in various parts of the universe and shewed thus that they arose from wise contrivance, I come next to consider the account which the atheists give of the origin of all things and examine its probability in comparison of what has been already advanced.

Lecture 79: The Inference to Design

Summary: Reid defends the argument from design, show-
ing that intelligence is inferred not by sense but from its
effects—just as wisdom or courage in others is judged
by conduct, not by direct perception. This inference, he
argues, is a first principle of reason—self-evident and
necessary, neither learned by experience nor derived from
logic. Experience shows conjunctions of observed things,
but wisdom, being unseen, cannot be learned so; rather,
we intuitively conclude that effects marked by order and
purpose imply a designing cause. Writers like Cicero and
Hutcheson illustrate the absurdity of attributing order
to chance. The argument grows stronger with scientific
discovery, as each law and harmony in nature magni-
fies evidence of intelligence. Reid reformulates it as
a logical inference: from marks of design we infer an
intelligent cause; such marks abound in nature; thus,
nature proceeds from wisdom. He rebuts Hume's claim
that singular effects preclude this inference by arguing
we know all intelligence, human or divine, only through
effects. Similarly, he rejects appeals to chance *(a con-*
fession of ignorance), necessity *(a chain of dependent*
causes needing a first cause), and nature *(a meaningless*

abstraction apart from a Lawgiver). Thus, the universe
cannot arise from blind forces but must originate in
an intelligent Creator—a conclusion strengthened, not
weakened, by modern philosophy.

Before I proceed directly to the subject proposed in the conclusion of my last lecture it may be proper to make some remarks
with regard to the argument which I last insisted on, viz that
from the marks of wisdom and design to be met with in the
Universe we infer it is the work of a wise and intelligent cause.
It is worthy of notice that intelligence, wisdom and skill are not
objects of our external senses, nor indeed objects of consciousness in any person but ourselves, and it may be observed that
even in ourselves we are properly speaking [not] conscious of
any either natural or acquired habit which we possess. We are
conscious only of their effects when they are exerted. A man's
wisdom can be known only by its effects, by the signs of it in
his conduct—his eloquence by the signs of it in his discourse.
In the same manner we judge of his courage and strength of
Mind and of all his other Virtues—it is only by their effects
that we can discern these qualities of his Mind.

Yet it may be observed that we judge of these [talents] with
as little hesitation as if they were objects of our senses. One
we pronounce to be a perfect idiot incapable of doing anything
that will be valid in law, another to have understanding and to
be accountable for his actions; one we pronounce to be open,
another cunning; one ignorant, another knowing. Every man
of common understanding forms such judgements of those he
converses with; he can no more avoid it, than he can seeing
objects that are placed before his eyes.

Yet in all these the talent is not *immediately* perceived, it
is discerned only by the effects it produces. From this it is
evident that it is no less a part of the human constitution to

judge of powers by their effects than of corporeal objects by the senses. We see that such judgements are common to all men and absolutely necessary in the affairs of life; now every judgement of this kind is only an application of that general rule, that from marks of intelligence and wisdom in effects, a wise and intelligent cause may be inferred. From the wise conduct, we infer wisdom in the cause, and from a brave conduct, we infer bravery; this we do with perfect security—it is done by all— they cannot avoid it. It is necessary too in the conduct of life; it is therefore to be received as a first principle.

Some however have thought that we learn this by reasoning or by experience. I apprehend it can be got from neither of these. We may observe that philosophers who can reason excellently on subjects that admit of reason, upon this subject appeal only to the Common sense of Mankind, and in some cases offer instances to make the absurdity of the opposite glaring and sometimes using the weapons of wit and raillery which in cases of this kind is very proper and often successful. Cicero[1] in his tract *De Natura Deorum,* speaks thus: "Can anything done by chance have all the marks of design? If a man throws dice and both turn up aces, if he should throw 400 times would chance throw up 400 aces? Colours thrown carelessly upon a canvas may have some rude appearance of a human face, but would they form a picture beautiful as the *Coan* Venus?[2] A Hog grubbing the Earth with his snout may

---

[1] Marcus Tullius Cicero (106–43 BC), Roman lawyer, politician, and philosopher; author of *De Natura Deorum* [*On the Nature of the Gods*] among many other works; the lines that Baird places in quotation marks do not correspond exactly to any passage from this text; however, Reid was apparently alluding to passages occurring in Book II at lines 88ff. and 93ff.

[2] The Coan Venus (also known as the "Coan Aphrodite") was a famous sculpture (not a painting, as might be thought from Reid's, or Baird's, choice of the word "picture") by the great Athenian sculptor, Praxiteles

turn up something like the letter A, but would he turn up the words of a complete sentence?"

Thus in order to shew the absurdity of supposing what has the marks of design could arise from chance, he gives a variety of examples where the absurdity is palpable without reasoning on the matter. And we find other authors arguing in the same way.

The ingenious Mr. [Hutcheson][3] endeavours to prove it by reasoning, and he is the only author I have met with who has made the attempt. Without pretending to say whether the reasoning is just or not I shall only observe, that he has drawn arguments from chances to shew that a regular arrangement of parts must proceed from design, that they could not proceed from chance. It may be remarked that this doctrine of chances is a branch of Mathematics not yet a hundred years old, but the truth of this principle has gained the assent of all since the beginning of the world and could therefore receive little strength from that reasoning.

Let us next consider whether it may not arise from experience, that from marks of design in effects we ascribe them to a designing cause. That this truth is not derived from experience is evident for two reasons. 1) Because this is a necessary truth and no Experience can discover a truth to be necessary. Thus, though it is consistent with our experience that twice three makes six, and the Sun always rises in the East and sets in the West, yet between these two all must perceive this distinction,

---

(c. 395–c. 330 BC). Cicero (see previous note) mentions the painting in the *De Natura Deorum,* but only in a different context, in Book I at line 75.

[3] Once again, we assume Reid intended Francis Hutcheson, though Baird again wrote "Hutchinson" (see note 1 of Lecture 73 above). In this instance, Reid may have been referring specifically to *Metaphysicae Synopsis,* Part III, Chapter I, Section 2. See Francis Hutcheson, *Logic, Metaphysics, and the Natural Sociability of Mankind,* edited by James Moore and Michael Silverthorne (Liberty Fund, 2006), pp. 152–153.

that the first is a necessary truth and has been and will continue true independent of our experience or of any cause, but the latter is not a necessary truth; it is contingent and depends on the will of the Maker of the World. If two things appear to us constantly conjoined and if our experience of this is uniform, this still gives no reason to conclude that they are necessarily connected or that they cannot be disjoined. In a word, Experience informs us only of what has been, not of what shall be.

Further, Experience may shew us a constant Conjunction between two things in those cases when both things are perceived, but if one only is perceived, experience never can shew it constantly conjoined with the other. For example, thought is connected with the thinking principle, but how do we know that thought may not exist without a Mind? These no doubt we find connected but if a man says he knows it by experience he deceives himself. Mind is not an object of Consciousness however—one only of them is perceived—we cannot say then that they are constantly conjoined. We conclude therefore that the necessary connection of thought and the thinking principle is not learned by Experience.

The same reasoning applies to the inference of design in a cause from marks of it in the effect. The one is an object of Consciousness but not the other. Experience then cannot shew a necessary connection. Thus it appears then that from marks of design and wisdom to infer intelligence in the cause is a first principle learned neither by reasoning nor by experience; it is self-evident and assented to by all men. It is on this principle then, that my argument is grounded. There are clear marks of wisdom and design in the formation and government of the world, must they not arise then from a designing cause? And this argument has this peculiar advantage that it gathers additional strength with every improvement in knowledge, every discovery in Philosophy. We are told of Alphonzo, a Moorish king,[4] that when his philosophers explained to him their notions of our planetary system, he said that he could have made a better one himself. But the system they gave him was not the work of God, it was the fiction of Men; but since the present new theory was introduced no one has presumed to shew how it could be better. Indeed when we attend to the marks of wisdom and intelligence that appear all around, every discovery proves a new hymn of praise to him who is the Creator and Governor of the world.

This argument has commonly been called the argument from *final causes* and we shall accordingly use it without enquiring into the propriety of it. If reduced to the form of a Syllogism these are the two premises: i) That an intelligent cause may be inferred from marks of wisdom in the effects; ii) There are clear marks of wisdom and design in the works of Nature. The conclusion is then—the works of Nature are effects of a designing and wise cause. Now it is evident that we must either deny the premises or admit the conclusion. The second of them I have endeavoured already at some length to prove and I have also shewed that the first is got neither by reasoning nor experience—that it is instinctive and believed by all men. But we find[5] among the Antients that the first of these was admitted, that those things which bear the marks of wisdom and design could proceed only from an intelligent cause and not from chance, but they denied that any evidence of these

---

[4] King Alfonso X of Castile (1221–1284); known as "Alfonso the Wise" for his support of all forms of scholarship, and also as "Alfonso the Astrologer" for his special interest in astronomy and astrology; the famous "Alfonsine Tables" were produced under his auspices and dedicated to him. Despite Reid/Baird's description of him as a "Moorish king," by this time in history Castile was back in Christian hands.

[5] The word "that" following here in manuscript has been suppressed as redundant.

were to be seen in the constitution of things. We may learn this from what we find put into the mouth of one of that atheistical sect in the third book of Cicero's *De Natura Deorum*.[6]

But modern improvements in philosophy have shewn the folly and weakness of this assertion and none now have the effrontery to deny that clear marks of wisdom are to be seen in the works of Creation. This appeared evident to the famous Galen who wrote a book *De Usu Partium*,[7] though he was an educated Epicurean,[8] purposely to shew that all could not proceed from chance. Those in modern times have seen the weakness of this and have that stronghold as untenable, but they have assaulted the other of the premises I mentioned, viz that we can infer design and wisdom in the cause from discovering it in the effects. In his dispute against this principle, Descartes,[9] though he surely was not an atheist, had led the way and his motive for it probably was this, that having invented some new arguments himself for the existence of the Deity, he wished to disparage all others in order to bring the greater credit to his own, or because he was offended perhaps with the Peripatetics for mixing final causes in their solution of the phenomena of Nature. A *Physical cause* is different from a *final cause*—the physical cause points out the laws of Nature from which the phenomena flow, thus, for example, we can shew that the physical cause of water rising in a pump is the weight of the atmosphere, but the *final cause* again points out the end which

---

[6] See note 1 of Lecture 79 above.

[7] See note 8 of Lecture 77 above.

[8] Follower of Epicurus; see note 8 of Lecture 73 above.

[9] René Descartes (1596–1650), French mathematician and philosopher. One of the most significant of the seventeenth-century authors who broke with the ancient and medieval tradition, he helped to usher in "modern philosophy" with such classic texts as *Discourse on Method* (1637) and *Meditations on First Philosophy* (1641).

Nature had in view. Thus, the end of the eye is for seeing, the foot for walking and so on.

These final causes Descartes thought he could not know; he thought the philosopher had nothing to do with them, and to attempt to explain them he considered as presumptuous and arrogant. In this trait he was not followed by many who admired him greatly in other things, particularly by the pious Dr. Henry More[10] of Cambridge and Fenelon,[11] Bishop of Cambray, who has wrote [sic] a book on the existence of God and his arguments are mostly drawn from the art of Nature, as he calls it, or those tokens of wisdom and design which appear in all parts of Nature. Since the time of Descartes however, we find that some have adopted his sentiments who may be suspected of a tendency to Atheism; of these we may reckon Maupertuis[12] and Buffon[13] but the most direct [attack] against this principle has been made by Mr. Hume,[14] who puts an argument against it in the mouth of an Epicurean[15] on which he seems to lay great stress—it is this, that the production of the Universe is a

[10] Henry More (1614–1687), English theologian and philosopher; author of *The Immortality of the* Soul (1659), *Enchiridion Ethicum* [*Manual of Ethics*] (1667), and *Divine Dialogues* (1668), among many other works; a leading light of the group known as the "Cambridge Platonists."

[11] François Fénelon (1651–1715), Archbishop of Cambrai; French theologian, philosopher, novelist, and poet; author of many works, notably *Dialogues of the Dead* (1712) and *Demonstration of the Existence of* God (1713); as Foster notes, Reid is most likely referring to the latter book.

[12] Pierre Louis Maupertuis (1698–1759), French mathematician, scientist, and philosopher; noted for his geographic observations which proved that the shape of the Earth is that of an oblate spheroid, and especially for his formulation of the principle of least action.

[13] Georges-Louis Leclerc, Comte de Buffon (1707–1788), French naturalist; author of the vast *Histoire Naturelle* [*Natural History*] in 36 volumes (1749–1788); he broached a number of ideas relating to heredity and "transformation" (evolution) a century before Charles Darwin (1809–1882).

[14] See note 11 in Lecture 73 above.

[15] Follower of Epicurus; see note 8 Lecture 73 above.

singular effect, to which there is no similar instance, therefore we can draw no conclusion from it, whether it is made by wisdom and intelligence or without. I shall consider a little the form of this objection.

The amount of it is this, that if we were accustomed to see worlds produced some by wisdom and others without it and saw always such worlds as ours produced by a wise case, the conclusion would then be this of ours was made by wise contrivance, but as we have no experience of this kind, therefore we can conclude nothing about the matter. This conclusion of his is built on this supposition of past experience, finding two things constantly united. But this I shewed to be a mistake. No man ever saw wisdom, and if he does not conclude from the marks of it, he can form no conclusions respecting anything of his fellow creatures. How should I know that any of this audience has understanding? It is only by the effects of it on their conduct and behaviour, and this leads me to suppose that such behaviour proceeds only from understanding. But, says Hume,[16] unless you know it by experience you know nothing of it. If this is the case I never could know it at all. Hence it appears that whoever maintains that there is no force in the argument from final cases, denies the existence of any intelligent being but himself. He has the same evidence for wisdom and intelligence in God as in a Father, a Brother or a friend. He infers it in both from its effects and these effects he discovers in the one as well as the other.

Having thus vindicated the argument from any exceptions that have been brought against it, I now proceed as I proposed to consider a little the causes which atheists assign for the Universe and the production of all this beautiful system.

---

[16] See note 11 of Lecture 73 above.

Some of them attribute it to Chance as the antient Epicureans[17] did, but we may observe that chance cannot be the cause of anything and when we say a thing happens by chance, it is a word expressive only of our own ignorance of the cause—chance can never be the efficient cause of anything. When a man throws a dice we say it is a chance which side turns up—now, what is the meaning of this? Chance can never be a cause; it means only that we are unable to discover the cause, for no man can measure the force with which he throws a dice with such accuracy as to tell what side will turn up, and it is so in all we attribute to chance. When a thousand tickets are put into a box and mixed and turned over by the motion of the lottery wheel, a boy puts in his hand and we say it is a chance whether he pulls out a prize or a blank—what is the meaning of this? It is that no man knows what it will be, whether the one or the other, but there is no part of it, but it has its cause.

To assign chance then as cause of anything is absurd and when the atheist tells you the Universe was produced by chance, he means only that it was produced he does not know how. This however does not hinder it from being owing to some cause, and that cause we have already shewn is eternal. Besides, nothing which we ascribe to chance is regular. If a man, says Tillotson,[18] throws down a heap of types on the ground it is a chance how they fall, but chance could not form them into a poem like the Iliad[19] or the Aeneid,[20] even into a discourse in Prose. Some

---

[17] Followers of Epicurus; see note 8 of Lecture 73 above.

[18] See note 5 of Lecture 75 above.

[19] The *Iliad* is a Greek epic poem recounting the events leading to the victory of the Greeks over the Trojans in the Trojan War; traditionally attributed to the semi-mythical, blind poet Homer. Originally handed down orally, the text is now thought by most scholars to have been written down sometime during the eighth century BC.

[20] The *Aeneid* is a Latin epic poem by the Roman poet Virgil [Publius Vergilius Maro] (70–19 BC); it recounts the flight of the Trojan hero

again have attributed all things to *Necessity*. But if we attend to the meaning of the words we will see Necessity cannot be the cause of any thing.

When we speak vaguely of causes indeed we say some of them are *necessary*, others Voluntary, though a cause in the philosophical sense of the word signifies only an agent by his own will producing the effect; yet in vulgar language we apply it to any instruments or means used to the production of the effect. So we say the pressure of the air is the cause of the Mercury rising in the Barometer and the necessary cause of this is the weight of the air which is as necessary as the effect it produces and this is produced by its gravitation, which also must have a cause and thus may we go on till we rise to the first cause of all. A necessary cause then is an effect produced by another cause till we are landed in a first cause of all, which is not necessary, but a real efficient cause. We shewed before the absurdity of supposing an infinite series of causes. I shall not now resume what was then said.

Another refers the Universe to *Nature*. They tell you Nature does so and so and that it produced all things. A late French philosopher has wrote a treatise in support of this doctrine, intitled *Systeme de la Nature*,[21] in which he ascribes everything to Nature. But what does he mean by Nature? It is a phrase as improper as *Chance* or *Necessity*. It is common indeed to say a thing is done by Nature to distinguish it from what is done by Art. Thus we say a Cabinet is the work of Art not of Nature —a tree again is the work of Nature not of Art. In this way

---

Aeneas from the ruins of Troy, eventually leading to the founding of the city of Rome.

   [21] *Système de la Nature* [*System of Nature*] was a rigorously materialistic treatise published anonymously in 1770; it created an enormous public scandal at the time. Paul-Henri Thiry, Baron d'Holbach (1723–1789), French man of letters and notorious atheist, subsequently claimed authorship.

we are accustomed to consider Nature as something opposite to art; we consider it as producing some things and art others.

But here by Nature we must mean it is either produced by the laws of Nature or by the author of Nature. Here the phrase is intelligible, but as an efficient cause it has no meaning. A law of Nature never could produce anything without an intelligent being to put them in execution. As in civil law it is not the law which tries a man, but the judge acting according to those laws and executing them. So the rules of grammar never would produce a finished creation of themselves without someone to form the sentences according to them. In like manner a law of Nature supposes a Lawgiver, a being who established and operates according to them. We see then it is vain to have recourse to this Subterfuge to say that all was produced by Nature—the term is as meaningless as if we said it was produced by Chance or Necessity.

Having now insisted so long on this argument I think it needless to insist at any length upon others. Some have argued the matter from the unanimous consent of all men in all ages except those who are so sunk in barbarity as hardly to merit the name of human creatures, and also—from this, that every part of the Creation bears marks of a recent formation. These and several others you will find in authors who have handled the subject, but though I would be very far from disparaging those arguments as useless, yet I would despair of convincing a man by these who resists the force of this—such a man is hardened beyond the power of arguments.

I now proceed to consider the Nature and attributes of the Deity.

Lecture 80: Reasoning About Divine Attributes

Summary: Reid urges epistemic humility: our concepts of God must be analogical and imperfect, derived from consciousness of our own mental powers and from observed effects in nature. He identifies three legitimate bases for reasoning about divine attributes: (1) inferring them from marks of wisdom and goodness in creation; (2) deducing consequences from God's necessary existence; and (3) ascribing perfections in the highest degree. Against Hume's restriction to the first basis and his call to proportion attributes strictly to observed effects, Reid maintains that unlimited perfections may be soundly inferred. He then treats the natural attributes. Eternity: by reflecting on duration—which has neither beginning nor end and is immeasurable—Reid concludes God's existence is coextensive with all duration. Necessary existence: unlike contingent beings whose being depends on a will, God cannot not be so dependent; this parallels the distinction between necessary and contingent truths. Immensity (omnipresence): though our notion of space is obscure, analogy from our own faculties and the excellence found in creation warrant affirming God's presence everywhere. He cites Newton's Scholium and Pope's couplet to underscore that philosophical and scientific reflection support these conceptions.

This is a subject too high to be grasped by our weak and limited capacities. When we consider attentively the works of Nature we see clear indications of power, wisdom, and goodness, yet we see still much remain into which we cannot penetrate and of which we must be forever ignorant. If this is true, then in this case, may it not be expected that our notions of the great author of all will be imperfect and inadequate as our notions of his works? The divine Nature indeed is a more proper object for the humble veneration of the pious heart than of curious disquisition to the most elevated Understanding. The pride of philosophy however has spurred on some to excogitate how the World might have been created and how governed.

It was from this intemperate desire of comprehending the laws of the Universe, that many among the antients excogitated or invented their Theogonies and Cosmogonies, and among the moderns too gave rise to those Theories of the Earth and of the Universal Government of things which appear indeed rather as the reveries of fanciful men than as truth. And I may venture to affirm that all these Theogonies and Cosmogonies which are not legitimately deduced from observation will always appear as unlike the works of God as the castles built by children, which the next moment they toss over with their foot, to the most regular and finished piece of architecture; and we have reason therefore to think that our notions of the attributes of deity will be as imperfect as the notions we can form of his works. This consideration then ought to make us diffident in the conceptions we form of the Divine Nature—we have no means of forming conceptions in any degree adequate to the object. The Supreme Being operates, before us and behind us, on our right hand and on our left, and even within us, but we see him not.

In rude ages, we see men generally ascribe to him a form like that of their own with organs and appetites similar to their own. When men improve in refinement and knowledge, a little reflection leads them to form more [perfect] ideas of deity as of a more spiritual nature. But though they do not ascribe to him a bodily figure, or the organs and appetite incident to [the] human Body, yet they find themselves under a necessity of assigning to him something analogous to the Human Mind, such as understanding, will, and moral character. All our original notions of Mind and its attributes are got by a consciousness of its operations in ourselves and so we can form no concept of any attribute in the Supreme Being to which there is not something analogous in ourselves. As a blind man can form no consciousness of Colour, or a deaf man of Sound, so neither can we form a notion of anything belonging to the divine mind of which we have no consciousness in our own. And perhaps there may be attributes belonging to the divine Nature of which we can form no more a conception than a blind man does of colours. It may be proper then, to point out the topics from which we commonly reason on this subject before we proceed to consider particularly his Nature and attributes and these I think may be reduced to three heads.

1) From the appearance of such attributes in the operations of Nature we may collect that they exist in the Deity. Thus, from the appearance of wisdom in the structure of things we may collect his wisdom and from the tokens or appearances of good contrivance—of such contrivance as tends to the good and happiness of his creature—we may collect his goodness. This is precisely like collecting the characters of Men from their ordinary conduct and it is indeed the chief source from which, by reason, we collect the attributes of the divine Nature—from the tenor of his conduct in the administration of things as far as they fall under our view.

2) We may reason with regard to some of the divine attributes, from his necessary existence. As it is evident that this is an attribute belonging to the deity, so it lays a foundation for our reasoning to some of his other attributes. This is a mode

of existence quite different from that of those beings that are merely contingent and perhaps draws [consequences] with it, which by no means result from *it*.[1]

3) In like manner we argue from his unlimited perfections. If it appears that we should ascribe to the deity every perfection in the highest degree, we may from hence deduce more accurate notions of some of his attributes.

These I think are the different topics from which authors have reasoned with regard to the attributes of deity, notwithstanding some have advanced it as a fixed point that we can reason only from the first of these topics, viz from the appearance of his attributes in the operations of Nature we may collect that they exist in him. This Mr. Hume[2] lays down as a first principle and thence draws it as a consequence, that we ought to ascribe to the Deity no higher degree of wisdom and goodness than what appears in the construction of his works and from this he endeavours to conclude that we ought to ascribe to him not those qualities in an unlimited degree, but only in a certain inferred limited proportion. The argument is grounded on this: it is to be observed that we can reason with respect to the Nature and attributes of Deity from no other topic but from the appearances of these attributes in the contrivance and governance of things.

Having premised these things I may remark that the attributes of the Supreme Being are commonly distinguished into two classes: his *natural* attributes and his *moral* attributes. I shall begin with the first. By natural attributes we understand those in which his will is not concerned. By the moral attributes, those which give direction to his will and conduct.

---

[1] Baird has apparently omitted something from this sentence. The simplest remedy would appear to be to insert "the atheist" between "and" and "perhaps." However, there is no textual basis for such an emendation.

[2] See note 11 of Lecture 73 above.

Of the Natural attributes of deity the first we take notice of is his *Eternity*—that is his being without beginning or end of existence, or as the Sacred writer prettily expresses it, "from everlasting to everlasting God."[3] Our notions of his eternity are derived from the notions we form of duration. We have an immediate and instinctive belief of duration in every act of Memory, for it is essential to this that it relate to something past and that some interval of duration has interposed between it and the present. Every man in every act of remembrance has a conception of duration, but there is something peculiar in this conception of duration—that though we can assign limits to our own duration, to the duration of every created being, yet we can assign none to duration itself.

We had a beginning, but duration did not begin with us. There was a time when we were not, but then duration was. So it is with every being created by God—put their beginning as far back as you please, still if they had a beginning, there must have been a time when they were not and therefore duration did not begin with it. We see then though the whole creation had a beginning, yet Duration could have none. Its nature will not permit us to believe that it had a beginning and neither can it have any end. We see no impossibility in supposing all created things annihilated in a moment and others created in their stead, but this would be impossible if duration did not continue to flow equally when no created being existed. Thus we see that Duration considered in itself is necessary—it had no beginning and will have no end and we cannot suppose it limited without a contradiction.

Things which have a beginning occupy only a small part which bear no more proportion to duration than finite to infinite. If we conceive a right line drawn out without beginning or

---

[3] Psalms 90:2.

end, any part of it that could be measured, by inches or feet or even miles would bear no proportion to the whole. Such a part must have a beginning and an end. It can have no proportion then or as arithmeticians say it cannot be an aliquot part of the whole which has neither beginning nor end.

Another thing remarkable in the Nature of Duration is, that as it is unlimited, so it is *necessarily existing*. We cannot say so of our own Selves. For though we have existed for a certain time, yet there is no absurdity in supposing that there was a time when we did not exist or when we will not exist. It depended on the will of God and he might give us existence or might not give it—it is a contingent event which may be or may not be. But can we say so of Absolute duration? By no means; it involves an absurdity in supposing there was a time in which it did not exist or will not exist.

Having premised these things with regard to Duration, I apprehend that there is no argument necessary to shew that the Deity is, was, and will be in all points of duration. His existence is commensurate with duration and there is no point of it in which he will not exist. And perhaps after all, this conception of the deity is inadequate and puerile. It is however the only one which our weak and limited faculties can reach. Some of the Schoolmen[4] thought to form a more adequate notion of duration by callingit "a moment continued forever without alteration" but this definition involves an absurdity in it, and instead of throwing light on the Subject, rather darkens it; for to suppose any point of duration to stand still involves a

---

[4] The medieval Latin philosophers based in the "schools," that is, the *scholae*, or new universities then springing up all around Europe; they included, notably, the great Italian philosopher Thomas Aquinas (1225–1274), based primarily at the University of Paris, and the English philosopher William of Ockham (c. 1287–1347), who trained at the University of Oxford before joining the papal court at Avignon.

contradiction and is inconsistent with every notion of duration which we can form.

It may be observed that all men, even atheists themselves, allow that something must have been Eternal and indeed this is evident from the first principle which I have already mentioned, that nothing can begin to exist without a cause; and for the same reason it will follow that what is uncaused and not produced by the power and wisdom of some other being must be eternal. Against any existence from eternity some objections have been brought, but they conclude equally against duration itself. The objections amount to this: That in the infinity of duration which is past, there must be a certain number of years, but there must [be] 12 twelve times as many months as years, and 365 times as many days as months, which is to make one infinity 12 times or 365 times as great as another. But this, if it shews anything, leads us [to] think duration had a beginning, which is absurd. The fallacy of the reasoning lies in this—that we apply years and months and days to measure that which from its very nature admits of no measure—infinity is immeasurable. Sometimes we apply infinite to express any large indeterminate number. In this sense it is intelligible, and in this sense, eternity may be said to contain an infinity of years, that is that no number of years can equal eternity. But if we use *infinite* to express any determinate number, then it is absurd and involves a contradiction no less than if we should talk of a square circle.

Another attribute of Deity is his necessary existence, that is, that it is impossible that he could not be. Every being that exists is either contingent or necessary. These two ways when opposed to each other are contradictory and therefore one or [the] other is applicable to every being. We call that *contingent* which either might or might not be, and that *necessary* which must be. Whatever either might or might not be depends on the will of some agent with power to bring it to pass or not—a power to produce evidently implies a power also not to produce —hence it follows that whatever be the cause of any existence, its existence is contingent and depends on the will of the agent whether it should exist or not. That we do exist is most certain —but that existence is contingent; the Supreme Being gave it and he can take it away when he pleases; it depends on his will.

But are we to suppose that the Supreme Being himself exists in the same manner? This evidently would be absurd; he derives his power and his existence from no other being. Therefore he is not contingent, but necessary. As necessary existence is a mode of existence of which no other Being but the Supreme Being is possessed, no wonder that it is too big for our weak faculties and that we find it difficult to conceive it. In order to assist our Conceptions we may illustrate the distinction of Necessary and Contingent Existence by the distinction between truths, which are also divided into Necessary and Contingent. Some truths we perceive manifestly to be necessary; such are all the truths of Mathematics, as this, that equals added to unequals makes the whole unequal etc. These we perceive are necessary; they must be true in all times and in all places without variation or change—they depend not for their truth on the will of any being.

But there are truths of another kind, which are contingent; thus the Sun has always continued to rise in the East and set in the West; this is a truth confirmed by uniform experience; nevertheless it is not a necessary truth; there is no contradiction in supposing the course of the Sun to be quite opposite, to rise in the West and to set in the East; it depends entirely on the Will of the Supreme Being that the one course takes place and not the other. Thus have we seen the distinction between Necessary and Contingent truths—that the one is true in all times and in all places and depends not on the will of any being, and that on the other hand, contingent truths depend on the Will and Power of the being who produced them. Now there is a similitude, though imperfect here to necessary and contingent existences. That is necessary existence which cannot not be, which must be and which depends not on the Will of any being. That again is contingent existence, which arose from a cause, which depends upon that power and may cease to be if that cause alters.

The next natural attribute of the Deity which I shall consider is his *Immensity*—that is, that he is everywhere present. Of a thing so far above us our conceptions must be inadequate. A child has only a dark and indistinct conception of those nobler powers in man which have not yet unfolded themselves in his infant heart; so our notions of the divine perfections and attributes must always be imperfect and puerile. Now we see them only darkly as through a glass and judge of them as children do of matters beyond their comprehension.[5] We can attribute nothing to deity of which there is not some faint ray or resemblance in ourselves. Whatever we perceive of real excellence in ourselves, in our own Mind, which is the work of God, it must be in the author of it. For no reasoning can be more convincing than this used by the Psalmist, "He that made the ear, shall he not hear? He that made the eye, shall he not see? He that gave man understanding, shall himself not understand?"[6] So if he has given man, the work of his hands, a certain sphere of action, shall not he himself act in a sphere far more extensive and circumscribed by no narrow bounds?

We have, however, I apprehend, a more distinct notion how body occupies Space than how the Mind does. By the sense of touch we receive certain notions of the places of Body; we know that two bodies can't occupy the same place at the same

---

[5] 1 Corinthians 13:12.

[6] Cf. Psalm 94:9–10.

time. We know too that Body is in its Nature finite, limited, and composed of parts totally independent of each other, yet Space is in its Nature unlimited and indivisible. We cannot call Space a Substance—neither can we call it relation with any modification of any substance—yet it is something of which we have a distinct conception. The subject is dark and intricate— there is something in the Nature of Space which overpowers and is sufficient to humble the most elevated Understandings when they see they cannot comprehend its Nature. Without then making any further reflections upon it myself I shall lay before [you] the sentiments of one of the most profound and penetrating geniuses ever the World produced, I mean Sir Isaac Newton. They are to be found in the Scholium annexed to his *Principia*, "*Deus aeternus est et infinitus, omnipraesens et omnisciens, etc. . . . cognoscimus.*"[7]—No human authority can be greater than this and they are the result of that deep reflection on the works of Nature, which makes him the lasting honour of his age and country. His great merit has been admirably and concisely displayed by Mr. Pope[8] in these two lines:

Nature and Nature's Laws lay hid in Night.
God said, *Let Newton be!* and all was Light.

---

[7] "We know . . . God is eternal and infinite, omnipresent and omniscient, etc." While Baird's Latin is somewhat garbled (we have regularized it here), the text quoted clearly refers to a passage in the "Scholium Generale," which Newton (see note 3 of Lecture 75 above) added to the second edition of the *Principia* (1713).

[8] "Epitaph Intended for Sir Isaac Newton in Westminster-Abbey," by the English poet, Alexander Pope (1688–1744); see Alexander Pope, *The Major Works,* edited by Pat Rogers (Oxford University Press, 2006), p. 242. Baird's manuscript departs slightly from the printed text.

Lecture 81: The Vastness of God

Summary: Reid continues his exposition of the divine attrib-
utes, focusing on God's immensity, omnipotence, perfection,
and omniscience. He begins by acknowledging human igno-
rance about how spirit can exist in a particular place,
affirming instead that unlimited space is filled by God's
immensity, just as duration is filled by his eternity. Two
arguments support this: (1) the evident presence and oper-
ation of divine power throughout all creation, and (2)
the necessity that a being existing of necessity must exist
everywhere. God's omnipotence is displayed in creation—
bringing matter into being ex nihilo—and in his continuous
governance of the universe through fixed laws. Reid rejects
Descartes' claim that omnipotence extends to contradictions
and insists that God cannot act against his moral nature.
Turning to perfection, Reid argues that every excellence
seen in creatures exists more fully in the Creator, whose
perfections are unlimited and absolute. Finally, he treats
omniscience, grounded in the pervasive marks of wisdom in
nature, in God's authorship of creation, and in the universal
human recognition of divine knowledge (evidenced by oaths).
Reid affirms God's knowledge of all things—past, present,
future, and even free actions—criticizing theories that reduce
foreknowledge to conjecture or "middle knowledge." Divine
knowledge, he concludes, is innate, unerring, and compre-
hensive, flowing necessarily from the divine nature.

We are at a loss in what sense to ascribe place to the Mind. The Schoolmen maintain that it was "*Totus in toto et totus in omni parte.*"[1] But this is so far from throwing light on the subject that it darkens it. We rather then should acknowledge our ignorance and confess this to be one of the many things beyond the reach of the human faculties. They must have a manner of existing in place totally unlike to body—a certain sphere in which they act and we cannot conceive action without an agent. Since then we find in ourselves a limited sphere of action and that we exist in a limited Space, it may [be] asked what sphere of action or what place can we ascribe to that being who is himself uncaused and the cause of all things, who is from everlasting to everlasting and who has unlimited duration with his existence and whose existence is no less necessary than endless duration and unlimited Space? Shall we consider such a one as confined by any limits? I conceive the most natural notion we can form is agreeable to that we have been endeavouring to establish, that unlimited Space is the sphere of his Power and that it is filled by his Immensity as duration is by his Eternity. But to reason more closely—the argument[s] which reason suggests on this Subject are chiefly two: 1) We see marks of his wisdom and power in all the parts of the Universe which fall under [our view]; 2) We may infer his Immensity from his *Necessary Existence.*

1) There are manifest indications of his power and presence through every part of his wide extended dominions. For all being was first created by an act of his will and power and surely could not be exerted when the agent was not. He laid the foundation of the Earth and the heavens, nay the heaven of heavens is the work of his hands. The most distant of the

---

[1] See note 4 of Lecture 80 above. The quoted phrase means "the whole in the whole and the whole in every part"; as usual, Baird's Latin is slightly garbled.

fixed stars also were made by him; now when we consider their distance, yet that their light though 12,000 miles distant are made visible to us, that they are in magnitude and lustre not inferior to our Sun and if we judge by Analogy, who are as distant from one another as from us and each having his *primaries* and *secondaries* revolving round him; if to this we add the immense number we see even by the naked eye and the still greater number by the telescopes, when, I say, we consider the Supreme Being as exerting his power and presence through this prodigious extent of Space, what measure can we set to his being, or to what limits of space can we suppose him confined?

Further: We are not to suppose the divine operation to cease, merely on giving existence to these objects. We see many changes and revolutions [in] Nature which requires the finger of Omnipotence to perform them, though to us they appear regulated by fixed laws. But we should observe, that the laws by which a being acts is one thing, and the being who acts agreeable to these laws is a different thing. It requires power and authority to act agreeable to laws, as well as to act without them. The being who acts without them we consider as possessed of no wisdom nor goodness, but he who acts by them we consider as possessed of wisdom and goodness and power nevertheless.

Now the Laws of Nature regular, constant, and uniform not only display his goodness and wisdom but require also his constant operation and therefore require his presence in all parts of duration. We formerly took notice of the Laws of Nature in the vegetable and animal kingdom as excellent and regular— in the manner of propagating their species, so various in the different kinds and so uniform in the individuals of each kind —in their manner of growth and of drawing nourishment—in the uniformity of their structure—and in a remarkable manner we perceived his operation in the instincts of animals and in the gradual evolutions of the powers of the Human Mind.

In the inanimate kingdom too there are many laws of Nature no less uniform and constant, as the cohesion of the particles of matter—their corpuscular attraction—various chemical affinities, all of [them] affecting only at small distances but affecting every particle of [matter]. There are other powers, as magnetism—it affects indeed at a greater distance but affects only one kind of iron matter. There are others still more extensive than this; thus, gravitation acts not only on every particle of matter on the surface of our Earth but to every part of the Solar System and by it every particle in that system acts on and is itself acted on by every other particle. Whether this power extends to the fixed stars we know not, but this we know, that the rays of light coming from them are subject to the same laws of refraction and reflection with the light in our own planets. The same laws of Nature thus operate through all uniformly and regularly.

2) There is an absurdity in supposing limits to a being, either in time or place, who exists *necessarily*. For this necessity is the same at all times and at all places and therefore we cannot suppose a being necessarily existing in one place which is not necessarily existing in another place. Though our notion of this kind of existence, as I observed before, is imperfect and inadequate, yet we can have a clearer conception of it by attending to the distinction of truths into *necessary* and *contingent*, and this is an evident property of truths which are necessary, that they are true in all places and at all times and could not possibly be otherwise. We conceive Space and Duration to exist necessarily, and therefore conceive them unlimited from their very Nature, and we can as easily conceive them not to exist as not to exist always and everywhere. What exists necessarily

then exists everywhere and at all times, having no relation to one thing or one place which it has [not]² to another.

Another attribute we have reason to ascribe to the deity is *Unlimited Power.* His power is manifested: i) In the works of creation. The power exerted in Creation is beyond our conception, for our human power consists in applying causes to effects, things active to things passive, but cannot produce a single particle of matter not before existing, nor can we annihilate a single particle that God has made. Indeed all the operations of Nature which we see consist in various combinations, compositions and decompositions of what is already made without either creation of new or annihilation of old. But it was not so from all Eternity—matter is a thing so imperfect that to ascribe an eternity of existence to it is absurd; all must be ascribed to the Deity, the great first cause of all.

ii) His power is manifested in governing the world. To suppose, with Leibnitz,³ that everything when created was endowed with such internal powers as to produce all the changes that afterwards happen in them is to exclude all divine management together. But we have no reason to think that the world is governed in this way, without the interposition of the Supreme Being. Nature leads us to conceive the Maker of the universe as its constant governor, and leads us to apply to him as the hearer of prayer and the kind protector of his rational offspring. The weak and imperfect power of man is

---

² The word "not" does not appear in Baird's manuscript here, but the sense clearly requires it.

³ Gottfried Wilhelm Leibniz (1646–1716), German mathematician, philosopher, and all-around polymath; author of *Essais de Théodicée* [*Essays on Theodicy*, or "Theodicy," for short] (1710), containing the classical formulation of the "optimistic" principle that God created the world in its present form because no better world was logically possible; according to this principle, as it was often put, this is "the best of all possible worlds" (with the emphasis on the word *possible*).

soon exhausted, but the power of the almighty is subject to no lassitude or fatigue.

We must not include however in our notion of his omnipotence the doing things impossible. This notion we find was advanced by Descartes[4] on purpose to support some parts of his system. He would not say, but that Deity could do things impossible; this is a mistake into which that philosopher [it is probable] would not have fallen had it not been that to support a favourite theory. For to do what is impossible is a contradiction in the Nature of things; we may as well conceive a thing to be and not to be at the same time. There are certain things which we perceive are necessarily true; now to suppose them subject to any power, even infinite power, is absurd for what is necessarily true is always true.

Another thing we cannot suppose the divine power to extend to is his *moral Nature*. It is absurd to suppose that he has the power of depriving himself of any of his perfections, as, his goodness, wisdom, justice, etc. It is no less absurd to suppose his power to extend to his moral Nature. When talking of a good or virtuous man, we say that it is impossible he should cheat or do an immoral thing; this expression is perhaps too strong when applied to Man, but it holds with respect to the Deity and is to be considered not as an impeachment of his power but an expression of his rectitude—not a defect in his Nature but a perfection of moral goodness.

Another attribute of the deity is *Unlimited Perfection*. This indeed is rather a general declaration of his attributes than of any one in particular, but it was necessary to take notice of it as from it we may argue some other of his perfections.

It may be observed that there are some notions of the Mind, which are general and abstract, yet are found in every

---

[4] See note 9 of Lecture 79 above.

individual of the species and that too very early; of this kind are our notions of *good* and *ill*, and what is nearly allied to these of *perfection* and *imperfection*. We give the name of perfection to that which every good man values in himself or in others, and what he wishes not to lose. Although men from different degrees of moral refinement may differ in their estimation of the value of things external, yet we find a very great unanimity in what qualities of the Mind are properly called excellences or perfections. Thus everybody agrees that ignorance and folly are imperfections, that knowledge and wisdom are perfections. All agree that power is a perfection and impotence an imperfection, that self-command is a perfection and being a slave to passion or appetite is an imperfection. To do what we know to be wrong and what we will afterwards repent and be sorry for, is an imperfection; to pursue steadily what is proper and right is a perfection.

Indeed I believe there is nothing in which men are more generally agreed, than in the application of these terms of perfection and imperfection to the qualities of the Mind. This shews us that these terms are not words without meaning, nor do they depend for their truth on the variable tastes of individuals, but that there is some common Standard by which they may be measured. Whatever implies defect, weakness, or disappointment or misery, we call imperfection; whatever is the object of esteem, love, veneration or admiration we consider as a perfection. Having thus laid down our notion of perfection and imperfection, what reason have we to ascribe to the Supreme Being every perfection in the highest degree?

1) It ought to be considered that every perfection or real excellence which we perceive in the creation belongs in a much higher degree to the Creator and perhaps in Deity there may be perfections of which we have no more a conception than a blind man of colours. If this is true however, as undoubtedly it

is, that every perfection in the effect is to be found in the cause, then we conclude that every real excellence we discover in God's Creation are only faint rays of more eminent perfections to be found in the Creator of all. But, 2) reason teaches us not only [to ascribe] to the Supreme Being perfection only in a superior degree but even in the highest degree. We his creatures are possessed only of that portion which he willed to bestow upon us—it is bounded by narrow limits. But what bounds can be set to his perfections, who is necessarily existent and who is unlimited?

Another natural attribute of Deity is *Perfect Knowledge* and *Wisdom*. The arguments reason suggests here are chiefly these two: 1) The marks of wisdom and design to be seen in the works of Creation. In all the parts of Nature, in water, air, earth and sea—in the disposition of things upon the Earth's Surface—in the wonderful variety of vegetables and the no less wonderful variety of animals—in the structure and instincts of animals— in the structure of the human body and Mind—in all of these, we found marks of a wise artificer, and the more we knew of his works, the more our admiration of his wisdom was raised. As some man ignorant of Clockwork, when he looks at the outside admires the regularity with which he observes the hours and minutes and seconds and can't help thinking that it required knowledge and contrivance to execute it, but when he is allowed to look into it and see the beauty and exquisite contrivance of the parts of the whole, how will his admiration of the skill of the maker be raised? It is so with our admiration of the skill of the great artist who formed the Universe. The various parts of the vast machine excite our surprise and wonder, but we are still more struck [by] the skill and wisdom which executed so stupendous a fabric. The wisdom and knowledge of others around us are discovered only by the signs of them in their conduct and actions; now we have the same evidence of the wisdom of the deity and in a much higher degree.

It may be observed: 2) The Supreme Being knows all his creatures and all their qualities, as it was him who first created them and who still governs them. The artificer knows his own workmanship. 3) We have reason to ascribe knowledge and wisdom to deity because it is a perfection. I formerly endeavoured to shew you that we have reason to ascribe every perfection to that being who is himself uncaused, independent and necessarily existing. Now no one can deny that knowledge is a perfection and a natural object of esteem, respect and love, and as it is found in some degree in the creature it must be ascribed in an unlimited degree to the creator. All the knowledge which man now possesses or ever will possess, nay, all that the most exalted *Seraphs* know is the gift of God; is it just reasoning then to say "he that giveth understanding shall himself not understand"?[5] Besides, in all ages and in all countries we find men disposed to ascribe such knowledge to the deity. This is manifest not only from writings of the heathens, but from the Universal use of an oath in all solemn transactions. Now why should they appeal to deity if they did not believe that he knew their actions? This is a clear proof of a belief of a deity who is the avenger of treachery and that from him nothing can be concealed. Indeed without knowledge and wisdom there could be nothing that deserves the name of perfection.

If we dare to hazard a conjecture according to our weaker faculties, concerning the objects of the knowledge of the deity, we much conclude he knows himself, his creatures and all their constitutions, that he knows all that has existed, that now exists, or that ever will exist; that he knows every relation of those things of which we now see only so small a part, and

---

[5] Cf. Psalm 94:9–10; Baird wrote "is it *not* just reasoning then," but the sense requires that the word "not" be omitted.

also, that he knows all the events of *necessary* causes, as well as of *free* agents. Some however have conceived that the future actions of free agents can't be known unless they flow from necessary causes but this never has been proved. We have indeed no notion of any such power in ourselves and are at a loss therefore to conceive it to belong to another, even to Deity, but this surely is no reason why it should not belong to him. It never can be shewn to be impossible why future free actions should not be foreseen and till this is done we have no reason but to ascribe it to the Supreme Being, and the cause of our being unwilling to allow it even to him is, that we do not possess it ourselves.

If this, however, is therefore a sufficient ground to deny it to him, why not also his creative power? We know how things already existing may be formed into various combinations, but how to give existence to what did not exist before we know not —are we therefore to deny that power to deity? Surely not. In the same manner we know not how the future actions of free agents can be foreknown and we know not how it is brought about, but we have this ground to think that it belongs to Deity, that *we* by *our* memory know actions that are past; yet there is no argument to shew the impossibility of foreseeing future actions which will not equally apply to Memory, and were it not that we ourselves are endowed with Memory, I apprehend, we would be apt to think that *it* would be as impossible as the foreknowledge of future actions.

Some other Philosophers unwilling to deny this power altogether with Deity attempted to account for it; but their attempts seem vain and impossible. Dr. Clarke[6] does it in this way: that we ourselves when we know the characters of Men,

---

[6] Samuel Clarke (1675–1729), English clergyman, theologian, and friend of Newton (see note 3 of Lecture 75 above); he entered into a correspondence with Leibniz (see note 3 of Lecture 81 above), defending

can form some conjecture how they will act in a given situation; so, as we must allow to the Supreme Being a more perfect acquaintance with the characters of men, hence he will have a more certain knowledge how they will act in any case. This is, I apprehend, reducing the divine prescience to an infinitely sagacious guess. Besides, it is supposing that men always act according to their characters, which is by no means true. Some of the Schoolmen,[7] particularly the Jesuits, in order to account for this foreknowledge, invented what they called a *Scientia Media*,[8] that is, not only a perfect knowledge of all that depends on necessary causes, but also of all events that can happen in any possible situation. With regard to their Scientia Media there were numberless disputes in the Schools, which it is [not] my design to enter upon here. We must also ascribe to the Supreme Being the possession of his knowledge without acquiring it by labour, exercise or slow degrees; a knowledge liable to no error, or Disappointment, but which is certain and unerring and which is the result of his own Nature and perfections.

---

the Newtonian position on a number of basic metaphysical assumptions underlying the *Principia*.

[7] See note 4 of Lecture 80 above.

[8] "Middle knowledge" is an idea introduced by the Spanish philosopher Luis de Molina (1535–1600) to reconcile God's foreknowledge of human actions with human free will; basically, middle knowledge is the knowledge God has of each person's character such that he knows hypothetically how each person would act in any possible situation.

Lecture 82: The Morality of God

Summary: Reid completes his discussion of God's natural attributes, turning to God's moral attributes. He emphasizes that divine knowledge is immediate, perfect, and all-encompassing, not gained by inference or experience, and that awareness of God's omniscience should shape moral conduct. God's spirituality signifies freedom from matter—immaterial, indivisible, and unconfined to place—contradicting all anthropomorphic or idolatrous conceptions. His unity follows from the uniformity of creation's laws and the impossibility of multiple infinite beings. God is also immutably happy, possessing perfect and self-sufficient blessedness. Turning to moral attributes, Reid argues that devotion must be grounded in confidence that God is not only great but good, and that his will is guided by moral perfection. Reason obliges us to ascribe to God every moral excellence—goodness, truth, justice —free from the limitations of human passions or instincts. The moral faculty in humans reveals genuine distinctions of right and wrong, reflecting God's own righteousness. Thus, divine goodness is manifest in creation's beneficial laws; his veracity is universally acknowledged as essential to faith in revelation and moral trust.

In all these natural attributes of Deity, though we find something analogous to them in the human Mind, yet we are to consider them as far removed from all that imperfection with which they are attended. In particular in the case of *knowledge*,

as formerly noticed, we are not to suppose the knowledge of
Deity acquired by slow degrees, or painful application, or by
inferring one thing from another; this is altogether inconsistent
with the perfection of the divine Nature. It must be supposed
to extend to all times, past, present or future, nay even to the
thoughts of our hearts, and it is not probable or conjectural
but certain and unerring. This consideration of the divine
knowledge ought surely to have a powerful influence on our
conduct. We know very well how great an influence the presence
of any character we respect has upon the human Mind. Much
more ought this to have, if we deeply considered and firmly
believed that we were always in the presence of God and that
our hearts are always open to him with whom we have to do.
It shews the folly of all hypocrisy and disguise as we cannot
be concealed from him with whom we are most concerned. It
ought to stifle also all vain glory and desire of men's applause
while at the same [time] it should render us careless of their
censure or contempt. Let us be more concerned to appear just
and innocent in his sight who is the best judge of our merit
and who will call us to account.

Another natural attribute of Deity of which we ought to
take notice is his *Spirituality*, which expresses not so properly
what it is, as what it is not. It signifies that his Nature is far
removed from Body or Matter, that it is not confined to place as
material things are; in short, that it has none of the qualities of
matter. We see however in men a great proneness to attribute a
visible form to the Deity, usually the human form, as thinking
it the most dignified with which they were acquainted. This
seems to be owing to the weakness of the human Mind, by which
we conceive of other things as like ourselves and we know and
judge by analogy. But of the absurdity of this there is sufficient
evidence, if we would consider with any degree of care that the
Supreme Being who is eternal, omnipresent, omniscient and

the first cause of all can never be material. All matter is finite from its very Nature; it is divided into parts and consists of a variety of parts each of which are distinct and independent of the rest and therefore is incapable of thought, for thought never can result from a composition of different beings. And as the operations of the Mind are one and indivisible, so this forms a strong argument that everything endowed with thought must be immaterial.

Though the evidence for this is clear and convincing yet have men conceived otherwise of the Supreme Being and such gross conceptions must be considered as the cause of that general spread of idolatry among the Heathens. Weak and foolish men conceived the Deity to inhabit only in their temples, but his presence is bounded by no such scanty limits for in *every* place will the prayer of the humble and acceptable worshipper be heard by him. Nay, they even imagined, that, if they worshipped on the tops of high mountains, their prayers would be better heard, as they were then nearer heaven than in the valleys. Strange! That reason should be so corrupted as to form such mean ideas of the great cause of all; yet so it is, that we see they were generally spread over the heathen world. But everyone who has rational notions of the Deity must conceive, that he dwells not alone in temples made with hands but that Universal Nature is his temple and that his ear is forever open to the cry of his Saints wherever they are placed. It is a natural and a just inference which our Saviour makes from this, that they who worship him must worship him in Spirit and Truth and that [it is] the homage of the humble and devout which is most acceptable to him.

Another attribute of Deity is his *Unity*, that is, he is *one* and not a number or plurality. We see how grossly the heathens erred in this, imagining all the Universe full of Deities, which was entirely owing to their mistaken notions of the divine

Nature. For had they carefully attended to the attributes which we have already run over, they might have seen that they could not belong to a plurality but to one. This may be argued from 1)[1] the form and uniform contrivance of the Universe. All appear under one government, subject to the same laws, and therefore there must be one Lawgiver. I had occasion to mention before a ray of light coming from the fixed Stars to our system, that is, from the most distant part of Nature which falls under our view, [is] governed by the same laws of refraction, reflection, and inflection, as those produced on the surface of the Earth or [our] own planets. We see the law of Gravitation by which various bodies on the surface of our Earth gravitate to the Earth, extends also to the Moon and not only to the Moon, but also to all the planets of our system, by which they gravitate to the Sun and to one another.

Thus from the most distant to the nearest parts of Nature all appear one great system, under one governor and subject to the same laws and not according to that absurd idea of the Heathen who conceived a vast variety of Gods, each of which had their separate departments, and that they often differed in their design and intentions, nay, sometimes quarreled and fought as represented in Homer.[2] But there is no such discord under the government of the Deity—the whole is under one governor and subject to [the] same laws. Besides, if we consider this a perfection which our reason tells us to ascribe to Deity, then it is impossible to conceive it as belonging to more beings than to one. We have already shewn that the Deity is eternal and immense and unlimited in all his perfections; now it is impossible to conceive a number of beings endowed with these. In all cases when we consider number, there must be something to distinguish the individuals of that number from each other;

---

[1] There is no corresponding "2" in Baird's manuscript.
[2] See note 19 of Lecture 79 above.

they must be distinguished by time or place or their Nature, or some other circumstance; take away all these, the plurality is lost and they really coincide with each other. This reasoning applies to the divine nature. We cannot suppose two beings endowed with those qualities to be distinguished either in time or place or by their nature or indeed any other circumstance.

I shall only take notice of another Natural attribute of Deity, viz, that he is *Immutably Happy*. Even the Epicureans[3] gave an eternity of happiness to their deities, but this they conceived to consist in an enjoyment of pleasure without any concern in the affairs of Mortals. This suited the tenets of Epicurus who placed their happiness in the enjoyment of Sense, but reason would lead us to consider it a perfection inherent in the divine Nature and that he who made all and is possessed of all power must be happy in himself. He is therefore called in Scripture the Blessed God, μακάριος, εὐλογητός,[4] meaning properly an object of praise or adoration. One cannot but take notice here from the account now given of the Natural attributes of God, of the perfect correspondence of what we find dictated by reason and the accounts given by the inspired writers. We see in the Old Testament, that Jehovah the Lord God, who revealed himself to the people of Israel, is everywhere represented as possessed of these qualities of Eternity, Immensity, Power, Perfection, Spirituality, etc., which from the light of reason we have now ascribed to him. And the same we find in the New Testament, and indeed it hardly can be supposed that such rational notions could be formed by a people so ignorant and gross as the Jews were and while the neighbouring nations were all deeply sunk in Idolatry.

Having said these things of the *Natural* attributes of God, I shall now consider his Moral attributes. There is no branch of

---

[3] Followers of Epicurus; see note 8 of Lecture 73 above.
[4] *Makarios, eulogētos*: both words roughly mean "blessed."

knowledge which it concerns us more to know than the moral attributes of the Supreme Being, for, on these depend all our hopes from him and from a knowledge of these will flow our behaviour toward him and the devotion we pay him. If we do not conceive the Supreme Being possessed of moral perfections in the highest degree, all our services to him will be the effect of fear and not of true devotion. True devotion can arise only from a belief that he is the best as well as greatest of beings and that our highest Virtue consists in resembling him as far as our weak faculties allow. By his moral attributes we mean those which relate to his actions and his conduct, by which his will and his operations are directed. He has an active as well as intellectual Nature.

The world was made by him and he upholds and governs it. And action consists in the exercise of power, and without the desire of acting, the power would be given to no purpose. Therefore to every creature to whom God has given power, he has given at the same time the principles of action to prompt them to the exercise of that power. Thus the instincts, appetites, and passions of animals all draw them on to action and the exercise of their power. But besides these incitements to action, man is possessed of a much nobler faculty—the moral faculty—by which he distinguishes right and wrong in conduct and distinguishes what he should pursue from what he should avoid. As by the Eye he perceives colours, by the Ear sound, and by the Memory past events, so by this faculty he perceives what is right and wrong—what is a subject of approbation, what of censure or indignation. This I shall have occasion to explain more fully afterwards. I shall there shew that the qualities of right and wrong which we perceive by the Moral faculty are really qualities inherent in the moral agent in which we conceive them to be and not, merely, as some hold, sensations in the percipient.

Now however we must take it for granted that there is a real and intrinsic difference between moral qualities, that gratitude, friendship, etc. are in their own Nature more worthy than perfidy, ingratitude, etc. Every man who consults his own breast must be convinced of this. If then there is a moral character in Man let us consider if we have reason to ascribe it to the Deity? Is there anything in the Deity analogous to moral character in Man? Here it ought to be observed that there are various principles of action in man which appear suited only to our dependent Nature and to [a] state attended with imperfection—these we never can suppose to belong to the Supreme Being. Thus we never possibly can ascribe to him, those instincts of which we see [common] in men and the lower animals, which lead them blindly to certain actions necessary for such imperfect creatures as we are, but to a being so perfect as the Supreme Being cannot belong. Neither can we ascribe to him those impulses which men receive from *Passions*; these were given to man to supply the defects of reason and the moral faculty. It is only what belongs to Man as a rational creature that we must ascribe to Deity. Now it is manifest that our Reason leads us to ascribe to him a perfect moral character. In order to confirm this I observe:

1) Every real excellence in the effect is to be found in the cause. No reasoning can be more forcible than this: "He that gave understanding shall not he understand?"[5] and the same reasoning leads us to attribute a perfect moral character to Deity from what we can discover in ourselves. He who made man capable of acquiring qualities worthy of esteem, respect and confidence, must surely possess these himself in an infinite degree free from all the imperfections which accompany them in our frail Nature. Besides, shall we ascribe to him knowledge

---

[5] Cf. Psalm 94:9–10.

and power and yet deny him righteousness and truth? Indeed there is no argument which can lead us to ascribe to him powers etc. which will not equally lead to ascribe to him those moral qualities which render the being possessed of them truly excellent and amiable.

2) Another reason why we should ascribe a perfect moral character to the Deity is from the moral government of the World. We see from the contrivance and administration of things that Virtue is countenanced and Vice discouraged. Virtue is in itself rewarded by the approbation of our own Minds; an approbation is felt from the practice of Virtue which forms the only sincere enjoyment attainable in this life. It inspires the mind with confidence in God and the hopes of a future reward; it affords a pleasure which never pales on reflection nor is ever followed by satiety or disgust. This can by no means be said of the other pleasures of our Nature. They are rather momentary in their duration and often on recollection yield ground for repentance and sorrow. Virtue again is countenanced by this that it tends to enlarge our own power in the world, to procure respect and good offices for ourselves and ours. Vice again is punished in the general administration of things, as in itself it is attended with remorse and dread of discovery—by the contempt of men and the lash of the civil law.

3) The Voice of Conscience leads us to ascribe a perfect moral character to the Deity. There is no sentiment more natural to man than this: Shall not the judge of all the Earth do right? I had occasion to observe before [that] trust in the Virtue of God was the firm support of injured Virtue and led all men to the expectation of a future state where a more perfect retribution would take place. This sentiment of the justice of the Supreme administration is surely what every man feels in himself. I hope there are none so bad, who commit any bad action without having some temptation to commit it, some

prospect of interest, some bodily appetite, or something or other which influences them; even the worst will do virtue when there is no temptation to the contrary. Now we cannot suppose the Supreme Being to have any temptation to do wrong—every creature is his—and all are in his hands; justly has an inspired Writer said, *"God tempts not any man neither is he tempted."*[6]

Having thus briefly shewed that we have reason to ascribe a perfect moral character to Deity, as well as natural attributes, I come now to consider what notion it is most reasonable to form of the nature of this moral character. Here I observe that the only notion we can form of his moral character is by ascribing to him what appears most excellent in ourselves separated from all the weakness and imperfection of human Nature. We ought to observe that there are some virtues grounded on our state as dependent creatures, liable to danger, error and misconduct, such as repentance, contrition, etc. These are suited to the state of Man, not to the state of the Supreme Being and can have no place in him who is exempted from error, danger, disappointment or mistakes. We are then only to ascribe to him such as imply no weakness or defect, such as: 1) Goodness, Mercy and Forbearance; 2) Truth and Veracity; 3) Regard to Virtue, and dislike to Vice; and 4) Justice and equity in the administration of things.

1) Goodness, Mercy and Forbearance are evidently implied in a perfect moral character, for without [them] we can conceive no moral character whatever. It appears from all the works of creation which are full of his goodness, that the laws by which the universe [is governed] are good, and indeed as far as we can know them, they are fitted to promote the interest of his creatures and to give all that degree of happiness of which their several natures are capable. These laws indeed are general and

---

[6] James 1:13.

sometimes through accident may produce pain, yet the state of man required the world to be managed by general Laws. If fire should sometimes burn and sometimes not, if water sometimes bears him and sometimes not, and so on of other parts of Nature, it would be impossible without the longest experience to acquire any prudence in our conduct. These general laws are necessary and are well constructed for the purposes of the various animals which are capable of happiness or misery. Thus the harms which attend any hurt are necessary calls to prevent us from neglecting to remedy what might endanger our health or our life. So the pains that follow the appetites of hunger and thirst are necessary to insure supply of meat and drink which we might otherwise forget. All these general laws serve as proof [of] the goodness of the Deity, as they evidently tend to the well-being of his creatures and the enjoyments of that happiness of which every nature is capable.

2) Another moral attribute of God is his *Truth* and *Veracity*. There is no attribute belonging to the Supreme Being, to which we more readily assent [than] this, and on which we more firmly rely. Some authors pretend to maintain that in reasoning with regard to the attributes of God, we ought to reason from no topic but from the appearances we observe in the Universe. Now it cannot be said that we have experience by means of our reason of his Truth and Veracity. They only have an experience of his Truth and Veracity to whom a revelation of his will has been made and who discern the Truth and Veracity in a conduct suited to that Revelation, but they who are left merely to reason can have no such experience; yet are all men found to believe in the Veracity of the Supreme Being.

Lecture 83: The God of Truth, Justice, and Love

Summary: Reid completes his treatment of God's moral attributes, emphasizing truth, justice, and love of virtue. Divine veracity is inseparable from perfect moral character— reason itself compels belief in God's truthfulness, even prior to revelation. God's justice is distributive, not commutative: he governs all impartially, rewarding and punishing in exact proportion to merit and guilt, allowing full weight to aggravating and mitigating circumstances beyond human judges' reach. Yet reason alone cannot specify punishment's measure or clemency's extent—these belong to divine wisdom. Reid then critiques skeptical systems denying God's moral perfections. Against Hume, who limits divine attributes to the degree seen in creation, Reid distinguishes intelligent causes from mechanical ones—human reasoning and common sense infer greater powers than those observed. Necessary existence, he argues, entails infinite perfections, not partial ones. Against Bolingbroke and Hobbes, who claim that moral attributes like goodness or justice are meaningless, Reid insists that religion collapses without them: moral imitation of God is the basis of virtue. The same rational grounds that compel belief in divine power and wisdom equally support belief in his righteousness and truth.

We daily experience his goodness, but of his Veracity we have no such experience; yet is the belief of his Veracity found to be inseparable from this a belief of his existence. Even human authority previous to experience has a weight with us, and let us suppose that the divine being should please to communicate anything to us by a revelation. I ask if any person could doubt of its truth? Is it possible anyone could think it all a lie? No man can entertain such a thought. And the reason of this seems to be that Truth and Veracity we conceive as inseparable from a perfect moral character.

3) Love to Virtue, and Dislike to Vice, is another moral attribute of the Deity. This is likewise in a perfect moral character for it is impossible to conceive a being endowed with any considerable degree of Virtue unless he regards the former and subdues the latter.

4) Justice. The writers on Jurisprudence have distinguished Justice into two kinds, *Commutative* and *Distributive*. The first regards our transactions with men when we consider ourselves as on a level with them, as in making a contract, carrying on traffic, etc. But the second is the justice of a Judge or governor in dispensing rewards and punishments in exact proportion to the merit and demerit of a person. Now if we ascribe Justice to the Deity at all, it evidently must be distributive Justice, that is, a disposition to deal with all his creatures without partiality or prejudice; making all allowances for those whom his Providence has placed in a more disadvantageous situation. And in the final distribution of rewards and punishments justice requires that he should not accept the person of any one in preference to another of equal merit and that every alleviation and every aggravation should have [its] full force; that the punishment should be proportioned to the degree of atrocity in the crime and at the same time that the rigour of punishment should be tempered by clemency in as far [as] Goodness will permit. This I think is the best notion we can form of the moral government of God and this is to be ascribed to the Supreme Being and it is that character which is everywhere attributed

to him in the Sacred Scriptures, where we are told that "he is no respecter of persons, but that in every nation he that fears him and does righteousness shall be accepted of him."[1]

But there are some peculiarities here and which ought to be attended to when we form a notion of the divine Nature. 1) Though it is a dictate both of reason and of conscience that an immoral conduct ought to be punished, yet they afford us no precept to determine the measure of that punishment. That criminal conduct deserves punishment is the voice of all men's conscience. "No one," says Plato,[2] "either of Gods or men, dares to say that Punishment is not due to the unjust."

... οὐδεὶς οὔτε θεῶν οὔτε ἀνθρώπων τολμᾷ λέγειν, ὡς οὐ τῷ γε ἀδικοῦντι δοτέον δίκην.

It is our indignation at this which makes all so ready to give their assistance in apprehending malefactors and in bringing them to punishment. In human courts, to be sure, the judge cannot pass sentence, as to the real demerit of the criminal; there is another rule which they use to observe: they are to judge how far the crime is hurtful to Society or Prejudicial to the interest of the Civil Government. They can judge by no other standard; they know not the heart, nor what measure of temptation he had or from what principles he acted. Of this

---

[1] Acts 10:34–35.

[2] Plato of Athens (428–348 BC), Greek philosopher; pupil of Socrates (see note 2 of Lecture 78 above) and teacher of Aristotle (see note 1 of Lecture 74 above); author of some 30 philosophical dialogues, most of which feature Socrates as the chief protagonist. The Greek passage cited just below, which in transliteration reads ". . . *oudeis oute theōn oute anthrōpōn tolmai legein, hōs ou tō ge adikounti doteon dikēn*," may be found in the *Euthyphro* at 8e1. The Greek found in Baird's manuscript departs considerably from the original; however, Reid's intended reference is clear inasmuch as Baird's Greek wording is roughly similar, and the meaning—as given by Reid's English translation—is identical. It seems likely that Baird has "back-translated" Reid's English rendering of Plato into schoolboy Greek. Thanks to Alex Jech for this suggestion.

the Supreme Being is the only judge, to him every heart is open and every circumstance that tends either to alleviate or to aggravate the guilt.

There are many crimes heinous in their own nature which a human judge cannot punish at all, as they are not absolutely necessary to preserve the peace of Society. Thus, ingratitude is in itself highly criminal, but it is not punishable by human laws if there is no injustice done; not that it is not considered as criminal but because it is not considered as necessary to the preservation of human Society. Xenophon[3] indeed tells us that under Cyrus,[4] the Persians by their laws punished ingratitude severely. By many however this account of the Persian constitution is considered as mostly fiction, and it is certainly true that never has any instance occurred in any well-regulated government where ingratitude was held an object of the Civil Law. Further, Reason does not dictate to us how far clemency should extend to penitent offenders. Justice certainly requires that an offender who repents and reforms ought to be otherwise dealt with than one who continues obstinate and impenitent, but how far this should be carried, whether they should be remitted altogether, or in what way or upon what terms they should be accepted, reason does not enable us to determine.

I have thus endeavoured to shew that we have reason to ascribe a perfect moral character to the Deity. I shall now consider some systems that have been advanced by authors of reputation concerning the attributes of the Deity and which contradict what I have now held forth as the fact.

---

[3] See note 1 of Lecture 78 above.

[4] Cyrus the Great (c. 600–530 BC), King of Persia (559–530 BC) and founder of the Achaemenid Empire, which repeatedly clashed with Greece. Foster suggests that Reid is alluding here to a passage in Xenophon's (see note 1 of Lecture 78 above) biography of Cyrus, the *Cyropaedia*, at I.ii.7.

1) I shall consider what has been said by Mr. Hume,[5] not indeed in his own person, but in that of an Epicurean[6] friend, whose sentiments he has held forth to us in his Essay on Providence and a Future State and adorned with all the strength of his reasoning and his eloquence, without either adopting or censuring it. He thinks that we have no reason [to ascribe] to the Supreme Being wisdom, power, or intelligence, in a higher degree than what we see Manifested in his works; a conclusion evidently grounded on this, that a cause is exactly proportional, to its effect; as therefore these marks of wisdom etc. are limited, so we must conclude that their cause, that is, the perfections of the Deity are limited. "When we infer any particular cause from an effect, we must proportion the one to the other, and can never be allowed to ascribe to the cause any qualities but what are exactly sufficient to produce the effect." It is allowed then that the Deity "possesses that degree of wisdom, power and benevolence" which appears in his workmanship, but "nothing further can be proved, except we call in the assistance of exaggeration and flattery to supply the defects of argument and reasoning."

I may observe here that this notion upon which the argument is grounded, that a cause is exactly proportioned to the effect, and limited to the effect, may perhaps be true of natural causes, but as to intelligent causes which operate freely and voluntary, this maxim is not founded on reason. I had occasion to observe formerly that this word *cause* is very ambiguous

---

[5] See note 11 of Lecture 73 above; the essay Reid refers to is "Of a Particular Providence and of a Future State," which is §XI of Hume's *An Enquiry Concerning Human Understanding* (1748). The first quoted passage may be found in David Hume, *Hume's Enquiries*, 2nd ed., edited by L.A. Selby-Bigge (Oxford University Press, 1902 [1972 impression]), in the editor's marginal section number 105, on p. 136; the second and third phrases in section number 106 on p. 137.

[6] Follower of Epicurus; see note 8 of Lecture 73 above.

—sometimes it signifies only some concomitant circumstances and sometimes even the Name of the cause is given to the Law of Nature itself. This is an important sense, but though from the use of Language we cannot avoid it, yet ought we to be cautious lest we be imposed on by the ambiguity of terms. In the proper and strict sense of the word we understand by cause, any agent with power to produce the effect and will to produce it. When we say cold is the cause of freezing in water, *cold* is here used in a vague and improper sense; cold is only a negation of heat and cannot be the cause of anything. But that is a cause of an effect which has power to produce it. We say too much heat is the cause of the liquor rising in the thermometer; here heat is used in a vague sense. But when we apply this maxim to intelligent beings, that the cause is proportional to the effect, it will be found to hold neither in reason nor in the common judgement of Men.

Suppose I should ask a man, on a journey, Pray, which is the road to Edinburgh? and he returns me a pertinent answer. Here the Understanding of the Man is the cause, the answer is the effect. Now, perhaps I never spoke to the Man before; I know nothing about him. Am I therefore to conclude that his understanding just enabled him to answer my question and neither more nor less? Surely this would be absurd—the natural conclusion is that he has such a degree, how much more I do not know. Again, if I converse with anyone half an hour upon Antient history and find that his knowledge is accurate, full and well digested, shall I say that this man knows no more than what I have heard him express? By no means; I am to conclude that he knows not only what he has expressed but much more. So it [is] with regard to his goodness. If I had stood in need of the benevolent assistance of a friend and that I found him always prompt to bestow his favours, do I conclude that he has merely that degree which he has manifested to me and that I

have exhausted all his Stock? It appears then that this maxim
of Mr. Hume's,[7] when applied to voluntary causes, is neither
self-evident nor consistent with our reasoning about causes in
common Life.

But still it may be said that we can consider the Supreme
Being as possessed only of that degree of skill, of power and
wisdom, etc. which we see displayed in his works. This indeed
is Mr. Hume's reasoning, but it is evidently grounded on the
Supposition that there is no argument for the perfections of
Deity except what is drawn [from] those indications of perfection which we see in his works. This no doubt is one topic from
which we do reason on the Subject, but it is not the only one,
as Hume supposes it to be.

I conceive that there is real force in our reasoning from the
Necessary Existence of Deity and his *unlimited Perfections*. It
has already been shewed by clear arguments that that being
who exists without a cause or a beginning exists necessarily,
so that it is impossible for him not to exist or not to have
such a degree of power, wisdom and goodness as is manifested
in his works. Now can it be said that necessary existence has
a connection with one degree of power which it has not with
another? When we consider a being possessed of necessary
existence we can see no connection he has with one portion
of time more than another, that therefore his duration is from
everlasting to everlasting. We conceive him to have no greater
connection with one part of Space than another, that therefore
he is omnipresent. In like manner when we conceive him as
endowed with one degree, we must consider [him] as possessed
of every degree, as there is no connection between necessary
existence and one degree which is not with another. This
reasoning applies to all the other attributes of God.

---

[7] See note 11 of Lecture 73 above.

I endeavoured to shew formerly that he was endowed with *unlimited Perfection*, from the consideration of his necessary existence. We cannot avoid ascribing different degrees of perfection to different objects; thus, we prefer a plant to a clod of earth, an animal to a plant, a rational to an irrational animal; and a being endowed with the highest degrees of perfection is the most perfect we can conceive. Now if there really is such a thing as perfection and imperfection we cannot help thinking that there is more perfection in the cause than in the effect, nor would the Deity have given us ideas of perfection beyond what he really himself possesses. We may observe that the reasoning of Mr. Hume's tends greatly to lessen the perfections of Deity— to reduce them from infinite to finite, and bring them on a level with our own, at least, to set them not so far above us as that a comparison may be drawn between the excellencies of Men and the excellency of God. Shocking thought! Presumptuous man! Does thou think with the short line of thy understanding to search the unfathomable wisdom of God? It is difficult indeed to say [whether] pride, impiety or presumption are most conspicuous in the man who makes the bold attempt; the man whose boasted understanding is unable to discover how one particle of matter adheres to another. Indeed the idea is so singular, that I once imagined that either Mr. Hume or his Epicurean[8] friend must have been the inventors of it, but I find that Milton,[9] long before Hume's time, has attributed it to Lucifer, who gives the same reason to encourage his associates in rebellion against their Maker but they were convinced of their error by this event.

---

[8] Follower of Epicurus; see note 8 of Lecture 73 above.

[9] John Milton (1608–1674), English poet and essayist; the two quoted verses are from his 1667 masterpiece, *Paradise Lost,* Book I, lines 92–94 and lines 143–145, respectively.

. . . so much the stronger proved
He with his thunder; and till then who knew
The force of those dire arms? . . .

. . . (whom I now
Of force believe Almighty, since no less
Than such could have o'erpowered such force as ours) . . .

I now proceed to consider another [topic] concerning the moral attributes of Deity, which has come from a different quarter but equally unfriendly to Religion and Virtue. I mean Lord Bolingbroke;[10] it had been advanced indeed by Mr. Hobbes[11] before and we find it adopted by Mr. Hume[12] in a Posthumous work of his on Natural Religion. He admits that there must be a first cause possessed of power, wisdom and the other natural attributes we have ascribed to him, but maintained that we know nothing of his moral attributes or the principles of his action; when we talk of his goodness, mercy, or justice, we use, says he, words without meaning. This system strikes at the root of all Religion, for if the Moral attributes of Deity are taken away, we can have no foundation for all the homage we pay him or for all the hopes we have from him. He who believes that the Lord is just in all his ways and holy in all his works, that justice and judgment are the habitations of his throne and that mercy and truth forever go before his face will of consequence believe

---

[10] Henry St. John, 1st Viscount Bolingbroke (1678–1751), English politician and philosopher; his copious writings tended towards rationalism and skepticism in matters of religion.

[11] Thomas Hobbes (1588–1679), English philosopher; in his masterpiece, *Leviathan* (1651), Hobbes adheres to a strict (if mostly implicit) materialism in his account of human nature and the origins of human society, making his name synonymous with religious skepticism, if not outright atheism.

[12] See note 11 of Lecture 73 above; the posthumous work referred to is Hume's *Dialogues Concerning Natural Religion,* published in 1779.

that real Virtue and real excellence consist solely in our resemblance to him, a consideration which gives them an authority they could not otherwise have. But this system cuts all the sinews of virtuous conduct, robs it of all its splendour and rests it on a very weak and slippery foundation. We certainly, however, have the same reason to ascribe justice and goodness to the Deity as power and intelligence nor is there the least ground to think his moral attributes more incomprehensible than his natural attributes. We acknowledge that our ideas of all the attributes and of the being of Deity are inadequate, but this is no reason why we should not have as distinct a notion of his goodness, justice and truth as of his power and intelligence. There are some vestiges of both in the human Mind and of the one our notions are as clear as of the other.

Lecture 84: The Problem of Theodicy

Summary: Reid defends moral realism and critiques "opti-mistic" theodicies. Moral distinctions are as certain as logical truths; therefore God's moral standard cannot invert ours. He examines the "Beltistan" (Leibnizian) view that God chose the best possible world and that all moral attributes reduce to benevolence aimed at maximizing total happiness (Bayes). Reid objects that a perfect moral character cannot be collapsed into benevolence alone and that this scheme tends toward fatal necessity, making evil an indispensable counterpart of good and undermining human freedom. Scrip-ture and reason warrant attributing multiple moral perfec-tions to God—love of virtue, hatred of vice, truthfulness, and justice—without compromising divine simplicity. Turning to evil, Reid distinguishes: (1) imperfections (mere lesser goods), (2) natural evils (pains that train prudence and fol-low from wise general laws), and (3) moral evils (free agents' misconduct). Objections from evil against divine goodness fail: imperfections are unavoidable in any world; natural evils serve ends under stable laws; the total balance of good and evil exceeds our view; and moral evil is properly men's doing, not God's—though God creates natures, sustains laws, and assigns lots, abuse of granted power is ours. He considers explanatory accounts of evil's origin next.

We can have no greater degree of testimony for the truth of anything than the testimony of our Senses and of these faculties

which God has given us; by these we discover that some propositions are true and others false and he who [has] a conviction
of the imperfection of his faculties [and] presumes to call in
question their information must remain a Sceptic forever; there
is no remedy for it. Now our judgement of right and wrong is as
certain as our judgement of true and false and to suppose that
the supreme being have another standard of measuring these
than we, that he thinks morally ill what we think morally good
and the contrary, is as absurd as to say that he has a different
conception of what is true and false. For it is no less evident
that goodness, justice, humanity, etc. are intrinsically better
than their contrary than that 2 and 2 make 4; it must surely be
allowed that a being of infinite understanding and intelligence
can discern these moral relations as well as we. If the difference
between moral right and wrong are distinguished by the human
Mind, much more so by him whose understanding is infinite.
It is an eternal and immutable truth that Virtue in its own
Nature is amiably respectable and deserving approbation; that
Vice on the contrary is an object of disapprobation, dislike and
demerit. When we judge so, we judge according to the truth;
the Supreme Being must be allowed always to judge according
[to] truth; therefore they must appear so to him as to the
human mind.

Having considered these two theories, which tend wholly
to overturn the attributes of the Deity, I now proceed to
consider some hypotheses whose authors in themselves were
not unfriendly to Religion, but which have been advanced
in order to render our notion of the divine attributes more
conceivable and to give such impressions of them as are most
agreeable to truth. The first of these I shall take notice of is
what is commonly called the *Beltistan* theory;[1] of this I shall

---

[1] The term "Beltistan" is rare and of uncertain origin. Both Duncan
and Foster are stumped by the word. Reid indicates it was in common

now give some short account and offer a few remarks upon it. Whoever wants it more fully explained may consult Leibnitz,[2] in his *Theodicè*,[3] who has adopted it. It has been also adopted by others, as that which gave the best account of the origin of Evil and the most amiable representations of the divine perfections and administration. According to this system, the Supreme Being from all eternity, by his infinite understanding, saw all the possible constitutions of [the] world which could be and their various qualities. Among all the possible systems that could be, he would chose that in which there was the greatest sum of happiness upon the whole. He then from his infinite understanding and his perfect goodness, constituted the present system as that which contained the greatest possible sum of happiness on the whole and that all the divine attributes

---

usage in his day, yet we have been able to trace only one earlier example of it, in the *Lectures on Divinity* (c. 1768) by the Scottish-American Presbyterian minister and educator, John Witherspoon; see *The Works of John Witherspoon, D.D.* (Edinburgh, 1815), vol. VIII, p. 108. Nevertheless, the meaning is clear enough. The Greek word "*beltistos*" is a superlative corresponding to "*agathos*" [good]. That is, it means "the best." Therefore, "Beltistan" is evidently compounded of this Greek root and the usual English adjectival suffix in its "-an" variant (Witherspoon writes "Beltistian," making the situation even clearer). Thus, "Beltistan theory" is etymologically analogous to the doctrines of "Optimism," or the "Rule of the Best," which occur in numerous earlier eighteenth-century texts and refer to the principle, associated especially with Leibniz's *Theodicy*, that we live in the "best of all possible words" (see note 3 of Lecture 81 above). Thanks to Cameron Wybrow for his help in solving this puzzle.

[2] See note 3 of Lecture 81 above.

[3] Baird's manuscript clearly shows a grave accent; this of course would be mistaken were "*Theodicè*" taken to refer to the French title of Leibniz's book (*Essais de Théodicée*). However, Baird may have intended "*Theodicè*" to refer to the neologism "*theodikē*," considered as a novel Greek technical term invented by Leibniz (by implication). In that case, the grave accent would be intended merely to distinguish the Greek letter "*eta*" from the letter "*epsilon*." (Note there is no accent, grave or acute, on the first "e.") Due to this possibility, we have let Baird's text stand as he wrote it.

consist in directing all things to produce the greatest degree of good on the whole.

This system leads [us] to form a particular notion of the divine attributes. It is conceived that though we give different names to the moral attributes of Deity, such as justice, truth, and righteousness, yet that they may all be resolved into one [and] are only different modifications of his goodness or benevolence; that therefore we have no reason to ascribe any moral attribute to Deity, but his benevolence, that is, a disposition to promote the greatest degree of happiness on the whole in the Universe. Some others since Leibnitz[4] have followed this system and particularly a divine among the Dissenters in England, whose name is commonly thought to be *Bayes*,[5] in a pamphlet entitled *Divine Benevolence*, in which he has endeavoured to shew that all the moral attributes we ascribe to Deity are only modifications of benevolence, or benevolence considered in particular lights. We see that according to this system, there is supposed in Deity no love of Virtue, or dislike of Vice, than as they tend to promote the happiness or misery of the beings in the world; that a desire of promoting the happiness of all is the only principle of his action and gave rise to his Laws and the government which he exercises.

By this system, it is thought the best account of the origin of Evil, both natural and moral, can be given. They think that all the Evil we see in the World is a necessary ingredient in a system in which we see the greatest possible good; it was proper then to admit it, and if we remove it an equal proportion of happiness is at the same time removed. We cannot help

---

[4] See note 3 of Lecture 81 above.

[5] Thomas Bayes (c. 1701–1761), English mathematician, philosopher, and Presbyterian minister; inventor of Bayes's Theorem and author of *Divine Benevolence: Or, An Attempt to Prove That the Principal End of the Divine Providence and Government Is the Happiness of His Creatures* (1731). Baird rendered the name as "Boyce."

thinking this a theory of the divine attributes, but it is a theory which, though well intended, has no sufficient arguments to enforce it, nor does it after all give us any clearer notions of the attributes of God than we had before. For, 1) As we can only form a just notion of moral character in Deity from what appears most perfect in moral characters among human creatures when separated from all the imperfections with which they are attended in us, so I Conceive that *goodness* alone is far from making a perfect moral character in Man. We cannot conceive a moral character without a regard to Virtue and a Dislike to Vice. To make the only principle of action in man to produce the happiness of others is to degrade his Nature. This, though a necessary branch of Virtue, is not the whole of it. There is no reason why the whole attributes of the Deity then should be resolved into one.

2) Though by this system we have the greatest possible sum of happiness, yet does it carry very uncomfortable prospects along with it and which appear very far from being agreeable to the Truth. For it supposes that in this system *evil* has a necessary and fatal connection with *good* and that it could not be removed even by divine Power. This is to suppose a Fate superior to the human being, which necessarily connects evil with the greatest possible sum of happiness. Likewise, we see that this system leads to the Necessity of all human actions, which indeed was maintained by Leibnitz[6] and the other patrons of this system, because it was necessary that every part should be so adjusted as to produce the greatest degree of happiness on the whole. Now if our authors affirm that the greatest possible sum of good could not be without that degree of evil we observe then they admit that there is a necessary and fatal connection between the one and the other. If however they

---

[6] See note 3 of Lecture 81 above.

do not adopt this system of a fatal necessity, if they admit these moral attributes, which we conceive as real perfections, then we have no reason to believe that evil is necessarily connected with good, nor is it necessary to reduce all his moral attributes to one class.

Perfect Virtue in Man consists not in a desire to promote the happiness of the Universe, without any regard to Truth, any love of Virtue, or dislike of Vice; now we form our notions of the Divine by the human character; if then we ascribe these [to] a perfect human character, they must be attributed equally to the Deity. Some again conceive that the attributing different moral attributes to the Deity is inconsistent with the simplicity and unity of his Nature which we ought to ascribe to an infinitely perfect being. But in this there is little force. Our conceptions of the Supreme Being are undoubtedly inadequate, but such as these notions are, they are the result of our faculties, and their imperfections must remain with us till our faculties are enlarged. We find too the Sacred Writers ascribing to Deity not only perfect goodness and benevolence but also of a being[7] of perfect purity who cannot behold iniquity; a God of truth in whom is no iniquity; these representations lead us to conceive of a moral character in the Supreme Being as we conceive it in a human Being but without the imperfections of humanity. Indeed if this were not the case, and if these attributes to which we give names in man had not the same meaning when we [ascribe] them to God, we would speak without understanding, and could reason no way with regard to them.

There have been others who through a love of a simplicity and to reduce the divine attributes to what they think consis-

---

[7] Baird appears to have left out one or more words here, but as the general sense is clear enough, it seemed to us fruitless to speculate what they might be.

tent with the Unity [of the] divine Nature, have included all under Divine rectitude and Divine Wisdom. But it does not appear that by reducing them all under one word we make the notion of them any clearer. Before I leave this subject we may consider a little the origin of Evil, a subject which has given rise to many theories, some of them absurd, and others which tend to darken it rather than throw light upon it. All the evil we see in the world may be considered in two different lights: 1) As giving rise to objections by some who are unfriendly to Religion, against a good government of the world and as a topic from which atheists draw arguments against a good adminis- tration of this world; 2) As a phenomenon giving occasion to the wit and ingenuity of philosophers and divines to exercise themselves in accounting for its origin and why [it is] permitted under a good government.

All evil has by some been reduced [to] 3 classes: 1) The evils of imperfection; 2) Evil which they call natural evil, 3) Moral evil.

1) By evil of imperfection no more is meant than this, that in the creatures we observe, there is not that degree of perfec- tion which they might have had; that is, that a man might have been much more perfect, he might have been an angel, a brute might have [been] a rational being and a plant might have been a brute animal. This however is not an evil, it is only a less degree of good. 2) There is *Natural evil*, that is, that suffering and pain which we see endured by beings in the Universe. 3) *Moral Evil*, that is, the violation of the laws of Virtue by moral and reasonable agents.

When this evil and Imperfection is offered as an objection against the good administration of things we ought to observe, 1) That objections which have equal force against any possible system which can be contrived have no force at all and are therefore to be rejected. Now with regard to the evil of imperfection, it appears impossible that any system can be made free from this objection. It therefore can have no force. Suppose a world twice, nay two thousand times more perfect than ours, still the objection remains, still they could have been more perfect.

Again, as to *natural evils*, if they are brought as an objection [against] a good administration of things, it may be observed [2)][8] that they answer many good ends. We see that it is by natural evil that men are trained up to wisdom and prudence in their conduct. Whether men could have been trained to that degree of Wisdom, Prudence and Virtue without these means we are not competent judges and cannot possibly determine, but from the present constitution of things we see they are necessary to our acquiring any prudence or wisdom, or patience or resignation. Besides that as far as we perceive, they are necessary consequences of good general laws. I shewed before that it was necessary for the constitution of rational creatures that they should be governed by general laws, for without these they never could pursue any means to the attainment of an end. And in a world governed by general laws occasionally evils will happen. If gravitation is a good general law and necessary to the preservation of our world, yet by this ruinous houses may fall and crush the inhabitants.

It may be observed, 3)[9] that we cannot determine what proportion this evil bears to the sum of the enjoyment of God's creatures. We see only a small part and can't judge of the whole of the Universe. If a man who was a stranger to Britain should land upon any corner of it and from that form an opinion of the whole, how uncertain must his Judgement be?

---

[8] The corresponding "2)" is missing from the manuscript.

[9] It is unclear why Baird has placed the number 3 here, instead of at the transition to the subject of moral evil, below. But then, his handling of the structure of Reid's lectures has been haphazard throughout.

Next, *Moral Evil*; this means the misconduct of [a] rational being and has also been objected to a good administration of all things by the Atheists. In order to judge of this, let us consider what properly can be said to be God's doing and what is not to be considered in that light. If on one hand we suppose man not to be a free agent, then every event good or bad is to be considered as God's doing, and the actions of the worst men are equally imputable to Deity as the rising or setting of the Sun. But if on the other hand, we suppose God to have given Man a Certain Sphere of Power, then the actions done in consequence of this are Men's doing and not God's. There is no maxim more evident than this that the action of one agent cannot be the action of another.

If men then are voluntary agents, no argument can be drawn from them either for or against the Supreme administration. What then can properly be said to be God's doing from which we may judge of his moral character? Here I observe, 1) Every creature is made by God and has its qualities from him; therefore these are to be ascribed to him. 2) What are the necessary consequences of that constitution are also properly his doing and operation. This is no less agreeable to reason than to the Sacred Scriptures. He made the Sun, Moon and Stars, he made a place for the Sea, he feeds the young Lions, and hears the savage cry and all of these he does either by means of some subordinate agent, or by his own immediate power. 3) To him we must ascribe the lot in which we are placed by his Providence with all its advantages and disadvantages. By such a connection with fellow men we are indeed liable to be sometimes hurt; this is a consequence of our situation. But such injurious actions are not to [be] attributed to God; he indeed gave the power, but they proceed from an abuse of that power. All moral evil then is not properly the doing of God but of men, who by abusing their power are liable to misery and are then justly punished for their misconduct.

It appears then that the objection against a good administration of all things, brought either from the evils of imperfection, natural evil, or moral evil, [has] no force. I now proceed to consider evil as affording room for exercising the wit and ingenuity of philosophers and divines to account for its origin and why it was permitted to exist in the world. This I shall speak of in my next Lecture.

Lecture 85: The Limits of Theodicy

Summary: Reid warns against speculative systems like the Beltistan theory, which presume to unveil the full plan of divine providence, explain the origin of evil, or reconcile every mystery of God's government. Such attempts, he argues, betray human pride, for even the simplest works of nature surpass our understanding. Just as one cannot infer the plan of the Iliad *from a few pages, so finite minds cannot grasp the vast design of the universe. Hypotheses framed to explain divine ends are as uncertain as competing "theories of the earth," each soon overturned by another. True wisdom rests content with the clear evidence of divine power, wisdom, and goodness without imagining comprehension of the whole. Reid contrasts this humility with the errors of pagan polytheism, which pictured gods finite and fallible, and instead commends the rational monotheism of the Jews as derived from revelation. He concludes by turning to the works of God, beginning with creation, affirming that the Supreme Being alone gives existence to what had none before and rejecting the ancient notion of eternal matter, since what is limited in place cannot exist necessarily.*

That the system called the *Beltistan*[1] was invented for the purpose of enabling us to comprehend more easily the moral attributes of Deity, the origin of evil, and the end for which it

---

[1] See note 1 of Lecture 84 above.

was permitted to prevail, is not to be doubted. But we must always judge that these authors who pretend to unravel all the mysteries of divine Providence and like Ariadne's thread,[2] pretend to lead us through all the turning and winding of this great Labyrinth, however much they deserve praise for their zeal, deserve little for their prudence or modesty. To comprehend the plan of the Universe and all the laws by which it is governed exceeds the utmost extent of human genius.

Presumptuous man that thou art! Wouldst thou vainly wish to be Privy Counsellor to the Almighty, thou who are unable to comprehend half the wisdom displayed in the meanest works of God? Consider the puny worm that crawls beneath thy feet and licks the dust of the Earth, doest thou know the end for which it was made, the useful purposes it serves to thee and to other animals? Canst thou unfold its structure? No man can; here the art of the skillfull anatomist is baffled; the Physiologist and the Philosopher are put to shame. Shall we then vainly attempt to comprehend the whole, who know not how one particle of matter adheres to another and how one body communicates motion to another? As soon may a mite comprehend the structure of an Orrery or unfold a system of Legislature, as we, the structure and plan of the Universe.

We see indeed in everything around us, in the curious structure of our Bodies and Minds, means excellently adapted to certain ends; we see a profusion of wisdom and power displayed,

---

[2] In Greek mythology, a creature half man and half bull called the Minotaur lived on the island of Crete. It inhabited an impenetrable maze called the Labyrinth. No one who entered the Labyrinth could escape from it. At this time the Athenians were required to provide human sacrifices to the Minotaur. However, a Cretan princess named Ariadne fell in love with Theseus, a prince of Athens. Ariadne convinced Theseus to slay the Minotaur, and to make his escape possible she gave him a ball of string. By unravelling the string on his way into the Labyrinth, Theseus was able to follow it back out again.

but all that falls under our view is only an inconsiderable part of the whole. Any man who should read a few pages of the Iliad of Homer[3] would have good reason to conclude that he was a very great Poet, but from such a small specimen no man that was not a fool would pretend to describe the plan of the whole or the manner in which it was conducted. In the like manner from the little we know of the works of God we have good reason to ascribe to him goodness, wisdom and power, but there is neither wisdom nor modesty in ourselves, when from the little we see we think to describe the plan of the whole.

In this respect the Beltistan theory[4] and all others formed to explain the ends of phenomena we see in the Universe may be compared to the various theories of the Earth which we have had by different authors. Many attempts have been made to explain the present appearances of things, of mountains, valleys, minerals, the different strata and layers of earths, those extraneous bodies, animal and vegetable, found at great depths in the Earth, and so on. Many ingenious authors have exercised their wit to invent a hypothesis to solve all these appearances. Accordingly, we find some attributing all to the universal Deluge, in which everything was displaced, torn up and tossed about and hence that mixture of marine bodies on the tops of mountains etc. which is to be found. Others again think the Mosaic deluge[5] insufficient for this purpose, and ascribe mountains and all these phenomena to the eruptions of Earthquakes and Volcanoes. Some again are of [the] opinion that the whole Earth was originally covered with Sea and as some places gradually wore away then others were left higher and became mountains. Others account for them by a gradual

---

[3] See note 19 of Lecture 79 above.

[4] See note 1 of Lecture 84 above.

[5] That is, the universal flood recounted in Genesis, 6–9; Moses was traditionally considered to be the author of the book of Genesis.

decrease of the waters, by which more and more ground was gradually left dry. Such are some of the conjectures about these appearances, and what do they amount to? They are only the dreams of Speculative men. Hence it is that every new theorist easily confutes the system of his predecessors and erects one equally flimsy in its stead, which falls also before them that come after him.

Now if this is the case in this instance, how can we expect to discover the plan of the Universe, of which we know so little and make so small a part? Hypotheses, indeed, of any kind, as I have often mentioned to you, brought to explain the appearances of things, are only the whims of a fanciful imagination and have always a higher probability of being found false and futile than true. As well may a child understand the various beauties or defects of the British constitution, or settle the balance of power in Europe, as we comprehend the plan of the great empire of God. In a word, so weak and contracted [is] our understanding that even in a human production, we meet with objections which we cannot answer and difficulties which we are unable to resolve.

In the meanest of Nature's works we are presented with difficulties which the short line of our understanding cannot sound. What presumption then is it, to attempt to discover the end, for whence the Universe was formed and all the different parts made subservient to the whole? In what we see no doubt we perceive manifest indications of wisdom and power and goodness from which we may conclude that these attributes belong to the author of all, but how ridiculous to think of comprehending the ends of the whole when we cannot comprehend the end of one of the meanest creatures. I observed before, that one from reading a few pages of the Iliad might have room to admire the abilities of the Poet, but surely he who pronounced from that with regard to the plan of the whole would justly be stigmatized as a fool. And if we cannot comprehend the works of Men how shall we pretend to comprehend the works of God?

In what we see there are proofs of wisdom and power, but all is far above our comprehension. He who takes it for granted that he is equal to the arduous task of unfolding the laws of the Universe will infallibly make blundering work of it and instead of solving objections by his theory, he will rather create more, and indeed we find that the objections of irreligious men are commonly made to these theories, which are not framed from a just observation of Nature, but are the creations of fancy and therefore more easily demolished.

Thus have I offered what I had to say with regard to the attributes of the Supreme Being and shall only make one observation upon the whole; and that is, that as the generality of men are little fitted for reasoning on subjects of this kind, as very strange perversions of sentiment concerning the truths of Natural Religion have prevailed among the heathens, though from what we have now seen, all things appear to be under one government and subject to the same laws, yet was the opinion of a plurality universal among them. The Immensity, eternity and unlimited Perfection of the Supreme Being, are necessarily connected with his necessary existence, yet the deities of the heathens were all conceived to have had a beginning, to have each his different departments assigned him, and to be limited with the follies and even vices of humanity. Such were their gross conceptions of their Gods, nor was all the philosophy of the polished Greeks and Romans able to root them out.

We must observe at the same time that [there] was one nation, I mean the Jews, who had more rational notions of the Supreme Being and of his attributes, notions perfectly agreeable to the dictates of reason as we have explained them. Now we can hardly suppose that a nation, barbarous as the Jews were, should have their reasoning powers more refined on this subject than their neighbours, without a Revelation from the Father of Lights. This then may teach us of how Great importance it is for [us] to attain proper notions of the Deity and his attributes, since so many of mankind have wandered so far from the truth and so contrary to what reason would dictate to them.

I now proceed to consider the Works of God, which is the last branch of this division of our Course. These have commonly by writers been referred to *two* heads: 1) The Creation of things; and, 2) His subsequent government of them.

1) With regard to *Creation*, that is, the giving existence to what had no existence before, we do not find this to have been the opinion of the heathen philosophers. We do not find that they conceived there was any such power as the giving existence to what had no existence before. They conceived that as there must be an eternal contriver and artificer of the Universe, so there must be an eternal matter necessarily existing of which all things were made. The Platonists[6] and Pythagoreans,[7] who are allowed to have formed the justest and most rational conceptions of the Deity, [...][8] yet we find that all of these sects maintain three eternal first principles: 1) An eternal cause, the Maker of all things; 2) An eternal matter of which all things were made; 3) An eternal idea or model according to which all things were made.

What seems to have led them to this error is that Creation is a work totally dissimilar to anything which is within the compass of human power. The Supreme being has given us the power of compounding and decompounding what already exists, but in us there is no vestige of the power of Creation;

---

[6] Followers of Plato; see note 2 of Lecture 83 above.

[7] Followers of Pythagoras; see note 1 of Lecture 75 above.

[8] Baird seems to have omitted text in between the two clauses of this sentence, which does not parse as it stands.

we in no instance can give being to what had it not before. All the works of human power extend neither to creation nor annihilation and their operation being so far beyond our own power we are unwilling therefore to allow it even to the Deity. But this is weak reasoning and if we duly exert our powers of reason, we will see it more probable that finite things must have their existence from the hand of the first cause of all. Indeed from the properties of matter it appears impossible that it should be eternal or necessarily existing. What exists necessarily must exist everywhere and in all points of duration but omnipresence and ubiquity belongs not to Matter; from its nature it is limited to one place; it cannot then be necessarily existing. And if we acknowledge that the Supreme Being gave existence to beings of a superior nature, even to rational Beings, it appears silly to attribute necessary existence to the meanest of all the creatures of God.

Lecture 86: Governance Divine and Human

Summary: Reid first rejects the notion that divine preserva-
tion is a continual act of recreation, arguing that such a view
would destroy personal identity and moral accountability.
Turning to the government of God, he distinguishes between
natural and moral governance. Against Leibniz's mechanistic
view of a self-sufficient universe, Reid contends that laws
require an agent to execute them and that divine oversight
is neither unphilosophical nor unworthy of God. In natural
government, God rules by general laws that are stable,
discoverable, and suited to human prudence, while instincts
and affections—such as parental love, social sympathy, imi-
tation, and even limited passions—are wisely implanted to
promote survival, learning, and virtue. Moral government, by
contrast, addresses rational beings capable of discerning right
from wrong and acting by choice. Humanity's present life is
a state of trial and discipline in which virtue is encouraged by
its natural rewards—health, respect, satisfaction, and hope of
future recompense—while vice brings disorder, suffering, and
contempt. In both natural and moral order, God's wisdom,
goodness, and justice are evident, yet Reid notes that only
minds freed from prejudice can rise, by reason, to proper
conceptions of the divine perfections.

When speaking of Creation we may take notice of a theory
which has been advanced by some Theologians, with a good
design no doubt, but which seems to draw dangerous conse-

quences along with it; it is this, that the preservation of God's creatures is a perpetual and constant recreation, and that therefore from the very nature of created beings, they must every moment fall into annihilation if not thus reproduced as it were. This is no doubt intended to represent more forcibly to us our entire dependence on the Supreme Being, but it must not be taken in too strict a sense; otherwise, our personal identity would be lost, for if what exists this moment is annihilation, then what exists the next moment is not the same with that which is now [no] more, nor can it be accountable for *its* actions.

Indeed it is impossible that we can form a notion of what is necessary to continue creatures in existence; we know not what it is as we know not the power which gave it being, but we cannot, I apprehend, reasonably conclude that it is the same with creation. A very ingenious Philosopher, as well as pious divine, gives us his sentiments on this subject. I mean Dr. Isaac Watts;[1] you will find them in his *Philosophical Essays*, Essay 11th, § 4, to which I refer you. What he says there is perfectly agreeable to reason and good sense, and I shall only add that this notion of preservation, being a continual creation, is destructive of all personal identity and of consequence of the accountableness of human actions.

I shall now consider the government of the Supreme Being which has been distinguished into two kinds, viz 1) His Natural government and 2) His Moral. On this subject we ought to speak with reserve and modesty and draw conclusions only from what our faculties are fully able to reach without pretending to form any conceptions with regard to the plan of the

---

[1] Isaac Watts (1674–1748), English Congregationalist minister, theologian, philosopher, and hymn-writer; author of a very popular *Logic* (1724), his *Philosophical Essays on Various Subjects* was first published in 1733; Essay XI of this volume is entitled "On Some Metaphysical Subjects," of which § 4 is "Of Creation and Conservation."

whole. According to the theory of Leibnitz[2] the world was so made as to need no operation of the Deity for its government; that everything had such power implanted in it at its first constitution, that produced all subsequent changes without any interposition of the Supreme Being and therefore he considered every interposition of the Deity as a miracle. This is a theory which has had many admirers but seems to have no foundation in truth or in reason.

It may be observed that he differs from the common meaning affixed to the word *miracle* when he considers every interposition of Deity as a miracle. It is not every interposition of Deity that constitutes an action *miraculous*; it is only action done in express violation of the usual fixed laws of Nature in order to attest a divine [omnipotence].[3] Thus the raising from the dead a man who has been four days in his grave and whose body is become putrid, by a single word, this is a miracle as it is contrary to the Laws of Nature. But that *every* interposition of Deity is a miracle cannot be admitted. We see indeed that the world is governed by general Laws, but do not laws require an agent to execute them and to produce effects according to them? Laws are not agents, they are only rules according to which an agent operates; the laws of Nature then suppose an agent to operate according to these laws, but whether the Supreme Being executes these immediately or by subordinate beings is beyond the reach of our comprehension.

Besides, it cannot even be shewn that this system of Leibnitz's is even possible. He endeavours to illustrate it from the structure of a clock. If, says he, a workman should make a clock that perpetually goes on of itself without needing any future interposition, any mending or reparation, this then surely

---

[2] See note 3 of Lecture 81 above.

[3] The sense seems to require "omnipotence" here, although Baird clearly wrote "omniscience."

would be a more perfect machine than one that required the hand of the artificer to be continually employed in regulating its motions and preventing it from going wrong. Now, all the works of God are surely perfect; the Universe then being the work of God must be perfect and therefore need no future interposition of his power to direct or support it. This similitude Leibnitz[4] relied on as very conclusive, but if examined carefully it will be found to lack much.

The workman indeed fashioned the materials and arranged them in a certain order but was he the author of these materials, or did he give them their powers by which the work is carried on? Did he give to matter that cohesive power between the particles which is necessary to the clock's being formed? Did he give it that tendency to descend or gravitate to the Earth by which the motion is caused and which, if it ceases, the machine must immediately stop? All that he does is only to apply certain powers, but it is nature and not him that confers these powers. Between these two then there is no similitude, neither is there a greater beauty in the system, than if we believed that all things are governed by a Supreme Being, or by some subordinate agent employed by him. Why should it be thought unworthy of Deity to preserve by his care, these creatures he formed at first by his power? Indeed, it is neither unsuitable to the principles of Philosophy or the Sacred Scriptures which everywhere represent him as the kind preserver of all his work.

With regard to the rules of the Natural government of God, it appears:

1) He governs all things by general Laws, as far as we can judge. We may however conceive the Supreme Being to have produced all things occasionally, but this is inconsistent with the moral government of his rational creatures. For were not all

---

[4] See note 3 of Lecture 81 above.

carried on by fixed laws *they* never could require any prudence, wisdom or forethought; it is by general laws therefore that the Supreme Being governs and in so doing his wisdom and goodness are conspicuous. These are called *Physical Laws*, in order to distinguish them from what are called *Moral Laws* by which is meant, those rules which ought to regulate the conduct of rational and moral agents. The former were appointed by the Almighty himself and are executed by him; they are therefore seldom violated and as I just now observed, were not the world governed thus, men never could acquire any prudence in their conduct through life, though he could as easily have done it by particular volitions.

2) These general laws which interest us most are made obvious to the experience of all and are soon perceived by all; thus, that fire will burn them, that water will drown, that bodies all gravitate. Such then as are necessary to be known are obvious to all; but there are others again more hidden which are left to the sagacity of man to investigate by his reason and industry. These which are commonly called Laws of Nature and which are the first principles of Natural Philosophy no more properly deserve that title than many of those laws which are obvious to the Vulgar; only the one is more hidden than the other. Those which are necessary to all are made obvious to all, but those again which are less necessary, though of use to enlarge human knowledge and human power are left to be discovered by our own sagacity and labour. These indeed have already gone great lengths; by them we have discovered why Planets roll in their orbits and the Comets are retained in their circles, and how far human genius may carry us no learned man [can] say. There is still ample room for the exercise of our talents from the beginning to the end of our existence and every new discovery we make tends to widen the sphere of our power and activity. For it is by means of our knowledge of the

Laws of Nature that we can bring about any end by using the properest means. It is by a knowledge of these laws relating to the fruits of the Earth that the husbandman knows when to plow and when to sow. By a knowledge of these the navigator totally traverses the wide Ocean, and, in a word, it is by a knowledge of these that every human art is carried on. The more then this knowledge is increased, the more will our power in these be enlarged. In the investigation of these, our talents find a manly and rational exercise and our labour seldom fails to be rewarded by the advantages that follow it.

3) In the government of God we see brutes and infants directed by other inferior principles which supply the want of reason; by instincts, appetites, and passions. Were not the child directed by instinct it must inevitably perish. Of itself it could never know that food was necessary to its preservation, far less that food was contained in the breast of the Mother and to be sucked out by its mouth; yet in all this, it is directed by instinct. Nothing shews more evidently the wisdom and superintendence of a Supreme Being than such instincts. We are unable indeed to discern the physical cause of them, but their effects we see, and find that they are admirably fitted for this purpose for which they were intended. Likewise, we may observe that men are directed by many inferior principles which supply the room of Reason and assist us in our progress in Virtue and moral goodness. That men were intended to make progress in Virtue is obvious, but the progress which is made by the greatest number is small and inconsiderable.

Society of consequence could not subsist if men were not directed by other principles of action to the same course to which Virtue itself would naturally lead them. We see thus that Society may subsist not only among those which are Virtuous, but even among those that are really bad, till they are corrupted in such a degree as we have few examples of in the history of the World. Men by means of their social passions, of natural affection, etc., though possessed of very little Virtue, are yet prompted to that very path which Virtue if followed would point out. It is the duty of the parent to rear, educate and protect their children, but were this done only from principles of Virtue, I am afraid that in the greatest number, we would find it neglected; to prevent this, we find implanted in every breast a natural affection, a στοργή,[5] as the Greeks called it, which operates on all and produces those effects which Virtue ought to produce. So in like manner the social affections, of gratitude, compassion, natural affection to relations and the Love of our Country operate equally on the good and the bad and supply the defects of Virtue. By these is Society supported and that whether the members of it are good or bad, virtuous or vicious, wise or foolish.

We see too principles implanted in Mankind which tend to our improvement in acts, knowledge and good habits. Of these we may take notice of that principle of *activity* in children which is instinctive and necessary for their acquiring habits for their improvement in knowledge. Even the perfect use of our Senses is to be acquired by habit and practice. Children thus by their desire of seeing every object with their eyes and handling with hands, improve them greatly and before they come of the years of understanding have them in perfection. I took an opportunity formerly also to shew that even our *perceptions* were mostly acquired and we see that Nature has fitted us with instincts fitted to acquire them.

Credulity too is evidently implanted in our Nature for our improvement. It is a natural law of our Nature, that even before we ourselves know the importance of knowledge, we listen with patience to what is told us and swallow it down with security,

---

[5] *Storgē*; a Greek word meaning love, especially the natural affection between parents and children.

by which means we acquire knowledge before we could learn it by our own discoveries. The *imitative* principle in Man was also intended for [that] progress in improvement. By this he is led to imitate what is done by others and thus easily acquires habits of great importance to him. We know from Experiments that have been made that it is even possible to teach deaf people to speak and pronounce articulate sounds though they have it not in their power to imitate these in others, by instructing them how to form the particular organs for the various sounds. This is no doubt a great art and requires great labour and attention. But the same difficulty would be found in everyone who learns articulate sounds and language, if it were not from the power of imitating these sounds by others.

We may observe also, that these principles given for self-defense, called *Malevolent Passions* though intended to promote our improvement and happiness in Society, yet from their very nature have checks to prevent their excess. They are attended with an uneasy feeling which is an admonition to indulge them no farther than what is necessary for our own good and the good of those to whom we wish well. We see by the constitution of things too that industry is encouraged as necessary to our subsistence. It is undoubtedly intended that Man should earn his bread by the sweat of his brows, that he should provide the necessities of life by his labour and for this we see him well adapted. And while we are prompted to action by the infamy, poverty and contempt which follows indolence and sloth, yet we are warned of the same by a languor and lassitude which attends too violent exertions of our powers to take alternate repose. From all these observations, we may remark that there are some obvious general rules of God's Natural Government which are admirably fitted to the condition of Man in this world.

Having said these things with regard to the *Natural Government*, I come now to the *Moral Government* of God. In the former he acts as a man does with his property; he disposes it in every particular as his wisdom and skill direct him; therefore whatever is done in the Natural world may properly be ascribed to God as his doing; such are the Motion of the Moon, of all the Planets, the ebbing and flowing of the Seas and so on. These all are the operations of the Deity, and the general rules according to which they are produced are called *Physical laws*; they are the rules to which he adheres and which of consequence are never transgressed. But in his *Moral Government*, he acts like a Legislator, who proposes rules of conduct to his subjects and as they obey or disobey them so may they expect his favour or displeasure.

As to the inanimate creation it is merely passive and can be subject to no laws. The brute animals again, though they possess a small degree of power and will, yet are they incapable of duty or of following any general rule of conduct; their actions are directed by blind impulse without being capable of distinguishing between right and wrong. The impulse which is strongest for the present always prevails and as that is the constitution of their Nature they cannot be blamed; they may be noxious, but they cannot be criminal; they may be objects of like and dislike, not of approbation and disapprobation. But with man it is otherwise.

Indeed, instinct and the blind impulse of our appetites and passions, as in the brute tribes, influence our actions in our infancy. By them we are governed, and therefore children and infants are not considered as capable of obeying Laws; they are not thought accountable for their actions; they cannot commit a crime. They are the subjects of discipline not of blame or disapprobation. But when they come to years of Understanding, they act from principles superior to appetites and passions;

they are capable of considering the consequences of actions. They can propose ends to themselves and prosecute them by proper means. They can reflect on a course of Life in others and observe the ends they pursue and consider the consequences of these pursuits. We can choose ends that are best on the whole. We blame ourselves when swayed by improper motives; we are led to do what we will repent of and wish undone. On the other [hand] nothing can give a Virtuous man more sincere or lasting joy than a consciousness of having preserved his rectitude unsullied in opposition to every powerful inducement to abandon it.

The consciousness of a wise and worthy conduct will always inspire with strength of Mind. It encourages the heart of a Man and makes his Countenance to shine, and the more costly the Sacrifice he has offered at the shrine of Virtue, he will find his triumph the greater. It appears then from this observation that there is one kind of action which we consider as deserving applause, another as unworthy. Nor is this distinction grounded on abstract disquisition but flows from the Nature of things. There is a right and a wrong, something worthy and deserving of approbation, though there were no human being to perceive it, and other things foolish, base and mean. This is the imme-diate dictate of our natural faculties as well as the distinction between true and false. If we form an idea of a man capable of approving perfidy, injustice, rapine, theft and so on and of disapproving of what was anyhow good or excellent, we would soon determine that his Judgement was as erroneous as if he should think twice three equal to fifteen.

Now the Supreme Being infinitely wise and intelligent, discerns human conduct in its truest colours and discovers whatever in it is worthy or blameable with all its aggravations and alleviations. His perfect moral character leads him to approve of what is truly worthy and to disapprove of what is improper. We must therefore conclude that God, in his government of rational creatures, has endowed them with the qualities of moral and rational agents.

Here we may observe that man is evidently placed here in a state of trial and probation, where he has access to improve in acts and knowledge, in Virtue and in good Habits. The present state is intended as a school of discipline and what we are to expect here is that proper means and inducements be set before us for our improvement, that proper incitements are held up to make us avoid Vice and pursue Virtue, as alone possessing real dignity and alone worthy of our approbation; of this a little reflection will satisfy us.

Here our condition is such that the good things we enjoy and every evil which we suffer are in some measure put in our power. By this I mean that a man by his foolish conduct may deprive himself of every enjoyment of life by his folly and imprudence. He may bring on himself cruel and tormenting pains that may shorten his days. He may reduce himself to poverty, disgrace and contempt and make himself the object of public hatred and of public vengeance. In the same manner all our good things are in some degree in our power as by our wrong conduct we may deprive ourselves of them. This then surely is a very strong inducement to look to our conduct as our enjoyments depend on it.

Further, our evils depend much on our conduct. I don't say *all* our evils, for the best men are designed to be trained to Virtue and happiness by suffering and trials, but of the common calamities of Life the greatest part are brought on by ourselves. For if we may trust to those who have made this subject their study, we will find that the greatest number of diseases are owing to our intemperance or to some wrong regimen; but health is generally enjoyed if we use temperance, proper exercise and a proper regimen. We see too that Industry is commonly able to furnish the necessaries and the conveniences of life; and if [by] Providence, they are ever reduced to want and indigence, they are entitled to our compassion and will also always find it; but poverty and all our ills are generally the consequence of idleness, intemperance and bad economy.

We see also how much our reputation depends on our conduct. If our conduct is worthy and irreproachable, whatever our station be, [we] will meet little respect from those who know us. High rank may display virtue in brighter colours and with greater splendour, but in the meanest also [virtue] will always be amiable and beloved. Thus we see that we are placed in a state of Discipline, so that our good enjoyments and even our [...][6] depend on our conduct.

But we may observe further that man is placed in such a condition that his conduct also has great influence on his fellow creatures. So has the divine wisdom seen fit to connect men in Society as he intended them to live in Society and mutually to assist each other. These circumstances interest us not only in our own conduct but in that of others; we are concerned that they should behave properly. Men therefore in such a situation have strong inducements to Virtue; indeed we may observe at the same time that the man who seriously intends to pursue a uniformly upright and worthy behaviour will find himself in a state of trial fitted for the exercise and improvement of his Virtue. And such is our constitution that Virtue is strengthened by exercise as well as our other habits are.

Such then the consequences of good or bad conduct. Great evils follow sloth, indolence, folly, etc. and that though not immediately, yet sometimes when the reason which caused it is forgotten, will it come. This is such an administration as might have been expected from the moral governor of the world.

---

[6] The ellipsis stands for the word "conduct," which is repeated shortly, but makes no sense here; what Baird meant to write is impossible to say.

The encouragements of Virtue and the discouragements of Vice are as strong as we can suppose in the present state. We are excited to Virtue by the tendency it has to create power, esteem and all the good and enjoyments of life, from the inward satisfaction we feel in doing our duty and the well-founded hopes of a future state. And as to the pains which even the virtuous sometimes are doomed to suffer, this is proper to a school of discipline and it is by these they are improved in every duty, so that [we] have reason to say with the antient servant of God, "*It is good for me that I have been afflicted.*"[7] And all their sorrows here will be abundantly compensated hereafter. Thus it appears that in the moral government of the Supreme Being, as well as in his Natural Government, he acts in a manner suitable to the perfections which by reason we were enabled to attribute to him.

We may observe at the same time that to form just notions of the Deity by mere force of our natural powers requires a greater impartiality and abstract research than is to be met with in the bulk of men. Accustomed only to the objects of Sense, though they may discover in the works of God evident marks of his being, power and wisdom, yet rude men, if left to trace out his attributes and perfections, will form conceptions gross and absurd, and far removed from the account I have now given you.

---

[7] Psalm 119:71.

Lecture 87: The Need to Think Rightly About God

Summary: Reid argues that while God's existence is evident from nature and providence, unaided reason in rude ages tends to distort the divine into polytheistic, anthropomorphic idolatries. By contrast, the Hebrew Scriptures uniquely preserved rational theism (one eternal, omniscient, omnipresent moral God) by revelation. He maintains that sound conceptions of the Deity are vital both personally and socially: they console individuals with filial trust in a wise, just, and benevolent Father and bind societies by restraining hidden crimes and deepening the bonds of justice and benevolence. Right theology also cures superstition—born of ignorance and false views of God—and allies with virtue, since true religion fortifies moral motives and provides sanction and example, whereas religion without virtue is hypocrisy and virtue without religion is often too weak against vice. Reid rebuts Shaftesbury's worry about "mercenary" motives, noting even Shaftesbury concedes that worthy views of God aid morals; hope of future recompense properly strengthens, rather than corrupts, virtue. Reid concludes by urging serious, candid reflection on these truths, commending Cicero's elegant testimony to an eternal, divine law ruling all.

Hence we find that the doctrines of Natural Religion have been also improved by the Speculations of Theologians and assisted by the representations of Deity given in the Sacred Scriptures. For nowhere do we find such a complete system of Natural Religion as in the Christian Writers. The being of God is indeed so evident from his works, and the conduct of his providence that no nation has been found so barbarous as to have *no* notion of Deity at all; yet it is to be expected that rude men, if left to trace out his attributes by the mere force of their reason, would form very gross conceptions, widely different from the representation of Scripture and the dictates of Sound reason. Mr. Hume,[1] in his Treatise on Natural Religion, has endeavoured to shew that men, especially in the early stages of society, with regard to their notions of Religion are prone to idolatry and to the conceiving a plurality of deities. And this no doubt is agreeable to fact as far as we have access to know, among every nation except the Jewish. Among all others the grossest notions have prevailed.

Familiarized with objects of sense, they formed ideas of deities with a human figure and with human passions. They conceived them as limited in their Nature and by no means everywhere present in the Universe. Each had his different department: one presided over the Sea, another the Earth, another the air and so on. They had also deities that belonged to every family, their *Lares*,[2] and these everyone worshipped by different rites. They imagined the different heavenly bodies too had different deities and to such extravagance was this spirit of polytheism carried among the Greeks and Romans that they had deities to every wood and grove and spring and river. But in all these notions there was nothing rational or that tended to

---

[1] See note 11 of Lecture 73 above; the "treatise" of Hume's that Reid refers to here appears to be "The Natural History of Religion," the first of *Four Dissertations* published in 1757.

[2] Originally, *lares* were guardian spirits in the ancient Roman religion; later, together with *penates*, they were considered "household gods," that is, gods protecting the home worshipped by the people in private.

improve the Human Mind. And it is highly probable that the enlightened writers on Morality among the Greeks and Romans left out Religion entirely from their system for this reason, that the notions of Deity publicly established were so absurd and so little suited to promote the practice of Virtue that they could expect no assistance from the principles of Religion in establishing the Principles of Morality.

Of all their 4 Cardinal Virtues, Prudence, Temperance, Justice and Fortitude, none have any relation to Religion or point out the Deity as an incitement to the practice of them. Nay, in some of these antient systems we find them maintaining that the deities interposed not in the affairs of men, and that, however powerful, yet they had no hand either in the framing or the government of the World. As this is the case, therefore that without the aid of Revelation our conception of Deity [is] low, it appears a strange phenomenon that the Jews, who were not more polished and civilized than others, but rather a barbarous people, should have such ideas of deity, his attributes and government as perfectly agreeable to what our reason dictates. They were not Polytheists; they believed in one God, the maker of the world, who was eternal, omnipresent, omniscient, who had a regard to Virtue and a dislike at Vice. Now such refined notions of Deity in a nation so rude as the Jews is hardly to be expected unless by a divine Revelation.

It is even probable that the notions of a polytheism at first arose from a Revelation, but were afterwards corrupted by the heathen nations. Idolatry was introduced and the veneration paid to men of worth and distinguished Virtue was converted into the worship of [them] as a divinity. Fables formed at first as pieces of moral instruction by degrees gained credit and were received as real stories. The just sentiments of the Deity were thus lost, by the corruptions of human reason, the craft of the priest or the cunning of the politician. We have seen that reason properly employed will point out the duties of Natural Religion, yet it is necessary to complete our notions of them, that we be enlightened by a divine revelation.

Having thus laid before you the evidence we have for the existence and attributes of the Supreme Being I cannot leave this subject without observing that it is of great importance, not only to the happiness of every individual, but to Society in general, to have just and rational notions of the Deity, his attributes, perfections and providence deeply impressed on the Mind. For,

1) There is no truth within the whole compass of human knowledge from which the Mind can derive such comfort. Is it agreeable to a child to know that he has a careful Father whose pleasure is to rear, educate, support and protect it? The Supreme Being is the Father of the Universe; the whole world is his care and his reasonable creatures are his children, so we find him represented in the Sacred writings and by an antient heathen philosopher as quoted in one of the Apostolical Epistles.[3] We are his offspring and our nature is obedient to his Commands. His wisdom and power were employed in our Creation, and still are in our preservation. He pities our weakness and infirmities as a Father and a Friend, and even to the wicked is he long suffering, patient, and ready to forgive. His ears are open to the cry of the young Lions, and much more to the humble and devout supplications of his rational creatures. In a word, all his administration is directed by perfect wisdom and with perfect justice.

---

[3] "Apostolic epistles" refers to the books of the New Testament consisting of letters written by St. Paul and others to various early Christian communities and individuals; however, as Foster points out, Reid seems to be referring rather to Paul's speech on the Areopagus in Athens, recounted in Acts 17:22–34.

And though we are unable to comprehend the unbounded scheme of infinite goodness, yet of this we are sure, that neither envy, nor jealousy, nor any malignant passion can disturb his happiness, or stain his perfections. At present, we cannot comprehend the plot of this great drama, and many scenes which shew the skill and intelligence of the great Poet may appear which we cannot account for, yet in this we may rest assured that at the conclusion every difficulty will be resolved and every incident shine forth as subservient to the design of the whole. And yet whatever difficulties we may meet with in this from the weakness of our faculties and the vast extent of the divine administration, yet a serious belief of the truths which our reason has pointed out cannot fail to fill every well-disposed Mind with confidence and joy. The Sun is not more necessary to the beauty and harmony of the Planetary System than the existence of the Father of the Universe to the comfort of every rational Mind. Let the atheist rejoice in the conviction of owing his being to a fortuitous dance of atoms, and let him rest in the uncertain hope of a future world by the same capricious fate! Surely the theist has much more solid ground to rejoice who considers himself as one of the offspring of God, who loves him and protects him. For he who trusts in God need have no other fear:

Je crains mon Dieu et je n'ai d'autre craint[e].[4]

2) A firm belief of the existence of Deity and of his providence is one of the strongest bonds of human Society. By means of his social affections we see that man was intended for Society and for mutually benefitting each other, but there

---

[4] From *Athalie* (1691), Act 1, Scene 1, by the classic French dramatist, Jean Racine (1639–1699). As usual, the citation is somewhat garbled; in the original, it is as follows: "*Je crains Dieu, cher Abner, et n'ai point d'autre crainte*" [I fear God, dear Abner, and have no other fear].

are some Men so wicked as to sacrifice all this to their lust of power or to some favourite passion. Now the government of God provides some checks to prevent these from going such lengths as they otherwise would do. The Contempt of good men [and] the Civil laws are strong restraints upon criminals; but those other crimes of which the laws have no cognizance —which are above the law—against these the belief of Deity and dread of a future Judgement are powerful guards. But the belief of these are not only powerful restraints upon the *worst*, but unites the *best* more firmly in Society. The Man who considers all as the children of the same Father, will find every tie of humanity, justice and benevolence strengthened by the consideration. And what more powerful incentive to promote the good and happiness of our fellow creatures than this, that in some degree we cooperate with the Almighty and merit his approbation, and that however our designs may be misconstrued by men, yet they will not be misrepresented by the great Judge of all the Earth. The atheist may complain that religion is the contrivance of the statesman to strengthen his laws and give stability to government; by this, he acknowledges that it is one of the strongest bonds of Society which could be contrived.

3) Just and rational sentiments of the Deity are of high importance as they guard us against Superstition. Two causes may be assigned for all the Superstition which has appeared in the world. i) Gross ignorance in the people which has emboldened cunning men to perform tricks among them. The success of this is always in proportion to the ignorance among the people to whom it is first divulged, but it has no connection with religion. ii) False notions of Deity, which have led men to believe that he is pleased with penances, and burying themselves in cloisters and sequestrating themselves from active life; such notions however can have no tendency to make men better

and the only remedy against them is the acquiring those notions which reason dictates to us and which Revelation confirms.

Lastly, they have a powerful influence in promoting and strengthening Virtue—True Religion and Virtue are natural allies and friends and cannot be disjoined without prejudice to both. Without a sense of Religion Virtue would of itself be too weak to restrain the vices of men and Religion without Virtue would be mere hypocrisy or black Superstition. The last has been allowed by all, but that Virtue without Religion is too weak to restrain the vices of men has been called in question by some, though, I apprehend, on insufficient grounds. Lord Shaftesbury[5] seems to be of [the] opinion that the inculcating the rewards and punishments of another life as an inducement to Virtue tends to promote a mercenary disposition, that the real and intrinsic excellence of Virtue is the only inducement that ought to be proposed. In this however his Lordship is not consistent, for we find in his Essay on Virtue and Morals that he takes up a contrary opinion, and acknowledges that rational sentiments of the Deity will have a tendency to promote the practice of Virtue in the World. Nor indeed can there be a truth more evident—will not he who believes the existence of Deity and that he delights in Virtue and abhors the worker of iniquity, endeavour to fullfill his pleasure and render himself agreeable to him by practicing what he approves and avoiding what is displeasing to him? The example of the Supreme Being he sets before him as his pattern after which to copy, and when the allurements of Vice are strong the consideration of futurity [is] called on to balance them; and if on the other hand, the

---

[5] Anthony Ashley Cooper, 3rd Earl of Shaftesbury (1671–1713), politician, philosopher, and man of letters; known as a superb stylist, his influential, three-volume *Characteristics of Men, Manners, Opinions, Times* (1711) collected a number of essays previously published by him. As a freethinker, Shaftesbury treated such subjects as natural religion, ethics, and aesthetics, among many others, in a mainly naturalistic manner.

incitements to Virtue are weak, the prospects of another world can add sufficient force to them.

Since, then, right notions of Deity are of such importance both to the individual and to Society, it becomes all of us to think of them seriously and candidly and endeavour to be established in a firm belief of them. I shall conclude all by recommending to your attention a passage in the works of Cicero,[6] *De Legibus*, Book 2, Chapter 4, where that enlightened philosopher hath expressed his sentiments on that subject with perspicuity and elegance.

---

[6] See note 1 of Lecture 79 above; Cicero's *De Legibus* [*On the Laws*] is a philosophical dialogue on law and society, in general, and on the relation between natural and positive law, in particular. In modern editions of *The Laws*, the three extant Books are not divided into "Chapters," but rather into much smaller sections, making it difficult to say with certainty which passage Reid has in mind here. However, the following speech from Book 2 by the character Marcus expresses the gist of Reid's ideas in this final lecture:

> This has, I know, been the opinion of the wisest men: that law was not thought up by human minds; that it is not some piece of legislation by popular assemblies; but it is something eternal which rules the entire universe through the wisdom of its commands and prohibitions. Therefore, they said, that first and final law is the mind of the god who compels or forbids all things by reason. From that cause, the law which the gods have given to the human race has rightly been praised: it is the reason and mind of a wise being, suited to command and prohibition. (Cicero, *On the Commonwealth and On the Laws*, edited by James E.G. Zetzel [Cambridge University Press, 1999], p. 132.)

Index

www.ingramcontent.com/pod-product-compliance
Lightning Source LLC
Chambersburg PA
CBHW071606210326
41597CB00019B/3429